Milestones in 150 Years of the Chemical Industry

Milestones in 150 Years of the Chemical Industry

Edited by

P.J.T. Morris
Science Museum, London

W.A. Campbell
University of Newcastle upon Tyne

H.L. Roberts
Formerly ICI, Mond Division

The Proceedings of a Symposium Organised by the Industrial Division of the Royal Society of Chemistry as part of the Annual Chemical Congress. Imperial College, London, 8–11 April 1991.

Special Publication No. 96

ISBN 0-85186-456-2

A catalogue record of this book is available from the British Library

Published by The Royal Society of Chemistry,
Thomas Graham House, Science Park, Cambridge CB4 4WF

Printed and bound in Great Britain by J.W. Arrowsmith Ltd., Bristol

Foreword

During the 150 years existence of the Royal Society of Chemistry, major contributions have been made to the quality of all our lives by the chemical industry. In order to celebrate the sesquicentenary of the Society and also these achievements, the Industrial Division decided, with the assistance of the Historical Group, to organise a symposium at the 150th Anniversary Congress. I was privileged to chair the Organising Committee which had as its members Drs Peter Morris, Alec Campbell and Hugh Roberts. Alfred Williamson was also an active member but sadly died just prior to the Congress.

To display the milestones of the whole of the chemical industry in one meeting was a daunting task. However, after due deliberation we found we could cover most, if not all, in six generic areas. These were, health, food, clothing, materials, energy and information. We were also able to assemble a distinguished array of speakers to present papers in their particular areas of expertise.

This book is the concrete record of that meeting and I am sure will be a reference book for those interested in the fascinating history of the chemical industry for many years to come.

I would like to thank the authors for providing their papers and to Peter Morris, Alec Campbell and Hugh Roberts for acting as editors.

Peter Bamfield,
ICI Specialties.

Contents

Clothing

Materials

Energy

Information

Dedication

This book is dedicated to the memory of
Alfred Gee Williamson

Introduction

Milestones in the History of the Chemical Industry

W.A. Campbell

DEPARTMENT OF CHEMISTRY, UNIVERSITY OF NEWCASTLE UPON TYNE, NEWCASTLE UPON TYNE, NE1 7RU, UK

Peter J.T. Morris

SCIENCE MUSEUM, EXHIBITION ROAD, LONDON SW7 2DD, UK

1 INTRODUCTION

After considerable discussion, the working party appointed by the Industrial Division to organise the 'Milestones in 150 Years of the Chemical Industry' symposium, chaired by Peter Bamfield, decided to abandon the traditional schema of the history of the chemical industry and to focus on the viewpoint of the final consumer, the man or woman in the street who benefits from the industry's products. Our aim was to show that the chemical industry fulfils a crucial role in the meeting of basic human needs, such as food or clothing, and to subordinate the details of chemical processes and the like to this end. This was a bold and fruitful decision, which has resulted in the 'user-friendly' format of the present volume.

When we came to edit the symposium's proceedings, it became clear to us that there was still a need to place the individual papers in an historical context that could not be met by short introductions to each section. Two of the editors, with expertise in different areas of the history of the chemical industry, therefore decided to put together an overview of the industry's history that would link the different papers and fill the inevitable gaps that arise in a volume of this kind. In particular, we were anxious to stress that the history of industry goes back well before 1841. Inevitably, this account is too brief to cover the rich and varied history of the British chemical industry in detail. The books listed in the short bibliography at the end of this paper should be consulted for further information. Nonetheless, we hope that this overview, read in conjunction with the papers that follow, will show that the chemical industry has often been of benefit to humanity, and that the history of the industry is not without interest.

2 THE EARLY CHEMICAL INDUSTRY

The early chemical industry made chiefly inorganic chemicals, had strong links with agriculture, and was almost entirely empirical. Bone ash, bone black, glue, grease, dried blood and Prussian blue were made from the waste products of animal husbandry. They served a number of trades ranging from pottery to paint, printing ink to fertilizers. The tallow that was too soft to make candles went to the soap-boilers. The by-products of forestry -- wood tar, wood spirit, bark, charcoal, turpentine, rosin and pyroligneous acid (impure ethanoic acid) -- were used in the manufacture of paint, varnish, mordants and gunpowder.

The copperas trade had operated on a small scale for over a century before it became a large scale industry in the 1750s. Pyrites (fool's gold or coal brasses) was hand picked from coal measures and stacked for several years to permit oxidation and hydrolysis. The product was green vitriol or copperas (iron (II) sulphate). It could be heated to form fuming sulphuric acid and a residue of Venetian red (iron (III) oxide). In a similar way, aluminous shale was weathered to produce an impure solution of aluminium sulphate, from which alum (the double sulphate of potassium and aluminium) could be crystallised after the addition of potassium chloride (from soapworks' waste). Both copperas and alum were used as mordants in dyeing.

Inorganic pigments were an important section of the early chemical industry. White lead had been made by the stack process since Tudor times; the replacement of animal dung by spent bark from the tan-yards took place in 1787. Red lead, made by roasting white lead, served not only the paint trade but also the manufacture of the highly lustrous lead crystal glass. Chrome yellow (lead (II) chromate) was first made by Andrew Kurtz, a German emigrant, in 1820. Prussian blue, discovered in Berlin in 1704, was manufactured after the 1750s by heating potash (potassium carbonate), scrap iron and slaughterhouse refuse. It was employed in the paint industry and by calico printers. Ultramarine, formerly made by grinding the expensive mineral lapis lazuli, was prepared in France and Germany in 1828 from china clay, soda, sulphur and charcoal. This is an example of the way the chemical industry was beginning to make available to a wider public substances that had been hitherto reserved for the very wealthy.

3 THE LEBLANC ERA, 1790-1870

The lead chamber process for sulphuric acid was developed by John Roebuck in 1746. Chlorine bleaching was introduced in 1787, and bleaching powder was first made by Charles Tennant in 1799. These three chemicals were crucial to the success of the Industrial Revolution and were linked by the Leblanc process for the manufacture of soda (sodium carbonate). The Leblanc process was developed by Nicholas Leblanc, the physician to the Duke of Orleans, between 1784 and 1790. He set up a factory at St Denis in 1791, but it was seized as part of the executed Duke's forfeited property three years later. The process was pioneered in Britain after 1815, most notably by William Losh of Walker-upon-Tyne and James Muspratt of Liverpool. The early Leblanc process was both wasteful and noisome. Five raw materials -- nitre (potassium nitrate), sulphur, salt, coal and limestone -- yielded only one saleable product, sodium carbonate. All the sulphur and nitre, and most of the chlorine were lost, resulting in polluted air and waterways, and the growth of evil-smelling wasteheaps.

The French chemist J. L. Gay-Lussac had invented a tower for the absorption of the escaping nitrogen oxides in 1827, but this was not widely taken up until John Glover introduced his denitrifying tower in 1859. The use of these towers in series brought the price of sulphuric acid down dramatically.

As early as 1836 William Gossage had developed a tower to trap the hydrogen chloride gas from the treatment of salt with sulphuric acid, but there was no large demand for the acid liquor which flowed from the tower. The passing of the Alkali Act of 1863 compelled the Leblanc soda manufacturers to adopt the Gossage tower, but the acid was often run into the nearest river. The introduction of Esparto grass into paper-making increased the demand for bleaching powder, for which purpose hydrochloric acid was converted into chlorine by treatment with manganese dioxide. It is no small irony that the chlorine bleaching of paper that is now roundly condemned by environmentalists helped to reduce the pollution of rivers by the Leblanc soda industry.

The recovery of sulphur from the evil-smelling 'galligu' waste of the Leblanc soda process was a more difficult problem, which was imperfectly tackled by several chemists including Ludwig Mond. The final breakthrough was made by Alexander M. Chance in 1885. He passed carbon dioxide through a slurry of waste and burnt the hydrogen sulphide thereby formed in the newly patented Claus oven to produce sulphur.

We have developed this theme at some length to show that while the chemical industry undoubtedly polluted the environment, it also spent much time and energy to find ways to reduce the pollution through waste recovery. In the 1860s, the British Leblanc soda industry was easily the largest in the world. Its production of over 270,000 tons of soda (ash and crystals) in 1866 demonstrates the large scale of the mid-Victorian chemical industry. However, within twenty-five years this chemical colossus was laid low by a wave of new chemical technologies.

4 DYES AND SOLVAY, 1870-1914

The first of these new technologies was not linked with the Leblanc soda process, but nonetheless helped to destroy the pre-eminence of the Leblanc manufacturers (and hence Britain) in the chemical industry. The founding of the coal tar dye industry by William H. Perkin in 1856 with his celebrated discovery of mauve brought much more than a new range of colours. The Leblanc soda industry could be, and often was, conducted by non-chemists. Indeed, mid-nineteenth century chemistry could throw little light on the reactions between semi-solid reactants, and analysis was too slow for effective process control. By contrast, dyestuff manufacture called for an understanding of and an ability to control a wide range of organic chemical reactions, including nitration, sulphonation, oxidation and reduction. A. W. Hofmann's research school at the Royal College of Chemistry provided entrepreneurial chemists who only carried out industrial research but often founded their own companies, such as Perkin himself, the partnership of Simpson, Maule and Nicholson, Greville Williams, and Carl Martius of AGFA. The scientific chemist had invaded industry.

The demand for coal tar products had consequences for the coking industry. Coke for metallurgical purposes was originally made in ovens which did not allow for by-product recovery, and the benzene, naphthalene and anthracene for dyes came from gasworks. From the late 1870s, coke ovens with facilities for the recovery of these by-products began to appear and the prejudice of iron masters against coke from the by-product ovens was gradually overcome. Coal tar distillation also underwent changes as new fractions assumed industrial importance. The demand for naphthols for azo dyes and phthalic anhydride for phthaleins led to a relative shift from benzene to naphthalene production.

Social changes followed in the wake of the new dyes. The advent of colours which could survive a day at the seaside benefited all classes of society. In any technological change there are winners and losers. In this case, the losers were the cultivators of dye plants, at home and abroad. However, most people shared Mr Punch's excitement about the new coal tar products:

'Oil and ointment and wax and wine
And the lovely colours called aniline,
You can make anything from a salve to a star,
(If you only know how) from black coal tar.'

The synthetic dyes were soon followed by the Solvay process for the manufacture of soda. It had been known since the 1820s that sodium bicarbonate (which is easily converted into soda on heating) would precipitate from a mixture of ammonium carbonate and saturated brine. This process was worked on a small scale by three firms in Britain and France between 1839 and the 1860s, but the chemical engineering of the day was unequal to the task of developing an industrial process that involved the handling of large quantities of gases and liquids. The loss of the expensive ammonia was a particularly difficult problem. The first viable process was developed by the Belgian Ernest Solvay in the mid-1860s. He licensed Ludwig Mond, another German emigrant to England, to work the process in Britain in 1872. Although they were beset with numerous serious technological and financial difficulties, Mond and his partner John Brunner were able to bring it to commercial success at Winnington in Cheshire.

Their cheaper, purer (and also more environment-friendly) soda posed a major threat to the Leblanc soda industry. However, the older process was able to survive for the time being because of the profitable manufacture of by-product bleaching powder. In a few decades, the Leblanc industry effectively shifted from being one single-product industry (soda) to another single-product industry (bleaching powder). Mond strove to recover the chlorine lost by the Solvay process, but he was not able to find an economic process. The later Leblanc industry laboured under the 'tyranny of stoichiometry'. In order to make the desirable bleaching powder, the chemical stoichiometry of the process compelled the Leblanc manufacturer to make a corresponding quantity of uncompetitive soda. The death-knell of the Leblanc process, then barely a century old, was sounded by the advent of electrolytic chloralkali processes in the 1880s and early 1890s. These processes used the newly available electric current from power stations to decompose salt (usually in the form of brine) into caustic soda (sodium hydroxide) and chlorine gas. The chlorine was cheaper than the Leblanc chlorine and caustic soda was more profitable than soda.

Even sulphuric acid manufacture was not untouched by technological change. Peregrine Phillips of Bristol had attempted to make sulphuric acid by direct combination of sulphur dioxide and oxygen over platinum as early as 1831. Little was known about catalysis and early efforts to develop this reaction usually failed through poisoning of the catalyst. Clemens Winkler of the Freiburg School of Mines in Saxony determined the need for pure reactants in the 1870s. Rudolph Messel, another German emigrant to England, applied the recently stated Law of Mass Action and used an excess of oxygen. He successfully operated the contact process at Stratford in East London in 1876. The lead chamber process survived for handling sulphur dioxide from pyrites, which could poison the contact catalyst because it often contained arsenic. The new strong 'contact' acid was particularly important for the dye industry which required pure sulphuric acid for sulphonation.

5 THE AGE OF MERGERS, 1914-1945

The decline of the British chemical industry relative to its German rival was made starkly clear by the outbreak of the First World War in August 1914. The most urgent need was for explosives, which created a demand for strong sulphuric acid which accelerated the shift from the lead chamber process to the contact process. Fortunately, the defeat of the German navy at the Battle of the Falkland Islands in 1914 enabled the supply line of Chili saltpetre (sodium nitrate) to be kept open, for Britain had no domestic method of fixing nitrogen from the air. The development of 'Amatol' (ammonium nitrate mixed with TNT) emphasised the need to solve the nitrogen problem. After the war, in 1920, Brunner Mond formed a subsidiary to manufacture ammonia using the well-known Haber-Bosch process at Billingham-on-Tees.

British production of synthetic dyes, already weakened by prewar competition from Germany, was further depleted by the need to produce picric acid (trinitrophenol) for filling shells. Read Holliday of Huddersfield was converted into British Dyes Ltd, and the rival Manchester firm of Ivan Levinstein (yet another German emigrant) took over the former Hoechst indigo factory at Ellesmere Port. At the end of the war, these two firms were brought together to form the British Dyestuffs Corporation (BDC) in order to meet the renewed threat of German competition.

Mergers were in the air. The Leblanc alkali manufacturers had combined in 1890-1 as the United Alkali Company (UAC) to make their last stand against the Solvay process, but the new firm had failed to appreciate the threat from the chloralkali process in time. What was left of the alkali industry outside the UAC had been mopped by Brunner Mond by 1918. In a similar manner, Nobel Explosives of Ardeer in Scotland had acquired most of the British explosives industry and several other industrial companies, and was converted into Nobel Industries in 1920. William H. Lever's Lever Brothers had taken over the entire British soap industry with the exception of Thomas Hedley of Newcastle and the Co-op. The next stage was triggered by events in Germany. The German dye firms had formed a tightly knit cartel ('community of interests') in 1916, which was converted into IG Farbenindustrie AG, the world's largest chemical company, in 1925. In the face of this threat, Nobel Industries pushed through the formation of a British counterpart, Imperial Chemical Industries (ICI), by the amalgamation of Nobels, Brunner Mond, the BDC and the almost defunct UAC in the following year. Lever Brothers merged with the Anglo-Dutch Margarine Union in 1929 to form Unilever Ltd/NV.

Many other mergers took place in the interwar years. Tar distillation, for instance, had been carried out by small gas companies, coal owners and coke manufacturers, as well as specialist distillation firms. After the First World War many of these companies began to amalgamate on a regional basis: South Western Tar

Distillers (1919); Midland Tar Distillers (1923); Lancashire Tar Distillers (1929); and Scottish Tar Distillers (1929). In 1920 several old established glue manufacturers were brought together to form British Glues and Chemicals. This merger activity peaked around 1930 with the formation of British Titan Products, British Paints, Goodlass Wall & Lead Industries, and the Imperial Smelting Corporation.

A few trade associations had been founded in the nineteenth century, including the Chemical Manure Manufacturers Association in 1875, and the Association of Tar Distillers in 1885. Nevertheless, the period following the First World War saw a rapid increase in the number of such bodies, beginning with the formation of the Association of British Chemical Manufacturers in 1916. This was followed by the National Sulphuric Acid Association (1919); the Rubber and Plastics Research Association (1919); British Sulphate of Ammonia Federation (1920, but descended from a committee formed in 1897); the Federation of Gelatine and Glue Manufacturers (1921); the British Chemical and Dyestuffs Traders' Association (1923); and the Association of British Manufacturers of Agricultural Chemicals (1928).

6 THE PETROCHEMICAL PERIOD, 1945-1973

At the end of the Second World War, the chemical industry was in the midst of a third wave of technological change. Plastics had begun to invade the lives of ordinary people in the interwar period, particularly in the form of Bakelite mouldings, such as the casing of the popular wireless sets. Celluloid was even older, dating from the 1870s, but had never managed to break out of its specialist niche in shirt collars and combs. Courtaulds had introduced viscose rayon in 1905 and the rival acetate silk was placed on the market by the Celanese company in the 1920s. Nonetheless, the story of modern plastics and synthetic fibres really began in the late 1930s with the development of polystyrene by IG Farben, nylon by Carothers at Du Pont in the United States, and polyethylene by ICI.

The Second World War saw the manufacture of synthetic materials, especially synthetic rubber, in the United States on a hitherto unimaginable scale. This scale of production, amounting to millions of tonnes a year, could not be sustained on the basis of coal as the major raw material. Fortunately, vast new sources of petroleum had been discovered in the Middle East in the

1930s, as well as natural gas in the United States, and as the price of coal climbed, the price of petroleum and natural gas fell.

The technology needed to convert these new raw materials into useful intermediates was supplied by the petrol-forming technology developed by the oil companies and independents such as Universal Oil Products and the Houdry Process Company in the interwar years. The petroleum (or natural gas) was cracked to furnish olefins, catalytically converted ('platformed') into aromatics, and oxidised to form fatty acids. The organic chemical industry had to shift its focus from the chemistry of coal-based starting materials such as acetylene and carbon monoxide to the petrochemical olefins, notably ethylene, propylene and butylene. Some the reactions used already existed, for instance the OXO-process developed by Otto Roelen of Ruhrchemie, which converted olefins into alcohols and aldehydes. Others, like the Wacker process for the conversion of ethylene into acetaldehyde (ethanal), were entirely new.

With the switch from coal to oil, the outward aspect of the chemical factory changed. Vats, furnaces, towers and chambers were replaced by the pipes and columns taken over from the petroleum refining industry. 'There is nothing to see' was a familiar reaction of 'lay' visitors to the new petrochemical factories of the 1950s and 1960s. Yet these unspectacular factories poured out vast quantities of organic chemicals, which in turn were converted into the resins, fibres, pesticides, solvents, dyes, and pharmaceuticals that lay at the heart of the 'chemical revolution' of the post-1945 period.

However, it should not be thought that technical progress was limited to organic chemicals in this period. The development of the atomic bomb, followed by the civil nuclear power programme, demanded a new technical chemistry of uranium, a metal hitherto used in small quantities in the glass industry. The introduction of the thermal diffusion method of separating U235 from the more abundant U238, via uranium hexafluoride, also stimulated the industrial development of fluorine chemistry. Titanium dioxide became increasingly important as a white pigment as health fears about white lead grew. Titanium itself was increasingly important through its use in jet engines and space crafts. New methods had to be developed for the manufacture of pure metals. This spilled over into the electronics industry, which began to demand ultra-pure silicon, germanium and boron. Boron also became a key element in the nuclear power industry as a moderator of neutrons in the nuclear reactor.

7 TOWARDS THE FUTURE, 1973-1991

Although vast technical changes had swept over the chemical industry in the forty years before the Oil Crisis of 1973, in corporate terms it was not very different. IG Farben had been broken up by the Allied Powers after the war, and its place in West Germany was taken by the three successor companies of BASF, Bayer and Hoechst. Nonetheless, the German 'Big Three' continue to be a powerful force in the worldwide chemical industry. The industry was still dominated by companies with strong historical roots, such as Du Pont or ICI, whose operations were largely confined to the chemical industry, and which were clearly identified with one country although clearly international in outlook.

The economic turbulence that followed the two 'oil shocks' of the 1970s did not lead to an abandonment, or even a curtailment, of petrochemicals. Petroleum and natural gas were still cheap compared with coal, and biotechnology has been slow to develop in the traditional areas of the chemical industry. However, the harsher economic climate after the halcyon days of the 1950s and 1960s gradually brought about a new pragmatism that was willing to sacrifice the links with the past in order to secure the future. The first major shift was Du Pont's takeover of the Continental Oil Company (Conoco) in 1981 and Seagram's shareholding in Du Pont as a relic of the fierce takeover battle. Another sea-change was the formation of transatlantic pharmaceutical corporations such as Smith Kline Beecham and Rhone Poulenc Rorer at the end of the 1980s. ICI's closure of its nitrogen activities at Billingham and the sale of its ammonia-soda business at Winnington to the newly formed Brunner Mond Holdings in 1991 represents that firm's most important break with the past since it closed down most of the remaining Leblanc works in the late 1920s.

If another history of the chemical industry is published as part of the bicentenary celebrations of the Royal Society of Chemistry in 2041, our successors may well regard the fifty years since 1991 as the most important -- if not the most traumatic -- of these two centuries. Clearly, the industry will have to meet and solve important challenges on several fronts including increasing concern about the environment, decreasing supplies of petroleum, possible technological shifts to coal-based and biotechnological chemicals, and the globalisation of the industry unfettered by the Cold War divisions of the last forty years. Whatever path the chemical industry decides to take, we are certain that the chemical industry of 2041 will be very different from the industry we know today.

SHORT BIBLIOGRAPHY

Campbell, W. A. 'The Chemical Industry'. Industrial Archaeology Series. London: Longman, 1971.

Campbell, W. A. 'Industrial Chemistry' in 'Recent Developments in the History of Chemistry', ed. Colin A. Russell, 238-252. London: Royal Society of Chemistry, 1985.

Clow, Archibald, and Nan L. Clow. 'The Chemical Revolution: A Contribution to Social Technology'. London: Batchworth, 1952.

Haber, L. F. 'The Chemical Industry During the Nineteenth Century: A Study of the Economic Aspects of Applied Chemistry in Europe and North America'. Oxford: Clarendon Press, 1969.

Haber, L. F. 'The Chemical Industry 1900-1930: International Growth and Technological Change'. Oxford: Clarendon Press, 1971.

Hardie, D. W. F., and J. Davidson Pratt. 'A History of the Modern British Chemical Industry'. Oxford: Pergamon, 1966.

Miall, Stephen. 'A History of the British Chemical Industry'. London: Benn, 1931.

Morgan, Gilbert T., and David D. Pratt. 'British Chemical Industry: Its Rise and Development'. London: Arnold, 1938.

Morris, Peter J. T., and Colin A. Russell. 'Archives of the British Chemical Industry, 1750-1914'. Faringdon: British Society for the History of Science, 1988.

Reader, W. J. 'Imperial Chemical Industries: A History'. 2 volumes. London: Oxford University press, 1970 and 1975. Volume 1: 'The Forerunners, 1870-1926'. Volume 2: 'The First Quarter-century, 1926-1952'.

Warren, Kenneth. 'Chemical Foundations: The Alkali Industry in Britain to 1926'. Oxford: Clarendon Press, 1980.

Health

The Contribution of Chemistry to Environmental Health

Sir Hugh Fish CBE

RED ROOFS, NEWBURY RD., SHEFFORD WOODLANDS, NEWBURY, BERKS
RG16 7AJ, UK

1 INTRODUCTION

Appropriately, it was some 150 years ago that environmental health first became a matter of public concern. It was essentially a product of the Industrial Revolution in the United Kingdom. Then the rapid expansion and crowding of towns to fill the workshops of the age of steam power shamefully fouled the urban streets, air and watercourses. The consequences of this evoked the great upsurge of Victorian zeal in public health reform and the creation of the great engineering works of water supply and sanitary drainage of the latter half of the nineteenth century. The British led the world in this broad reform, in which chemistry played a vital, and continuously increasing part.

In the modern context, chemistry in relation to environmental health is a very broad subject, embracing of course chemical engineering and applied biochemistry, and many aspects of environmental health. Only a part of this broad field of endeavour can be considered within the time span allocated to this lecture. It is, therefore, not surprising that I should choose to restrict my observations to the contribution of chemistry to environmental health in the water management context, in which I have been closely involved for most of my working life.

2 CHEMISTRY IN THE PURIFICATION OF PUBLIC WATER SUPPLY

The major sources of water supply are taken from natural lakes and artificial impoundments of upland rainfall run-off, from lowland rivers and pumped storages filled from rivers, and from chalk, limestone and sandstone groundwater aquifers. Generally the waters taken from the large and usually deep aquifers are naturally of the best hygienic quality. The upland waters in remote areas are also normally of good quality, whereas the waters taken from lowland rivers and shallow local groundwater sources are usually subject to various kinds of pollution, including bacteria and other organisms which are pathogenic to man. However, in the UK and in other highly developed countries, there are very few water sources which can nowadays be regarded as fit for human consumption without prior disinfection. In the underdeveloped countries, the typical village stream or well source of drinking water is exposed to contamination, directly or through the soil, with human faecal pollution and the common pathogenic organisms of human waterborne disease. Fortunately, this lamentable state of affairs in the underdeveloped countries is being steadily improved by the efforts of international institutions and charities, notably the UK organisation known as Water Aid.

In essence, the traditional treatment of water for public supply has three primary objectives. First, to remove suspended solids and colloidal matter to render the water substantially clear and colourless. Second, to disinfect the water to destroy any pathogenic organisms surviving the clarification treatment. Third, to ensure, through the proper application of the clarification and disinfection processes, that the water in delivery through the public supply mains remains satisfactory for public use. It is a well established fact, although occasionally forgotten or neglected, that these three objectives are of equal importance and all should be attained continuously, and that failure to achieve any one objective can cause the water drawn at the tap to be unsatisfactory. Nevertheless, the failure to achieve proper disinfection of water leads usually, but not always, to the most serious consequences to environmental health.

Chemical Disinfection of Public Water Supply

Over the period 1830 to 1865, calamitous outbreaks of cholera occurred in England. In 1854 there were 20,000 deaths from cholera, and the most famous outbreak in that year was at Broad Street pump which was investigated by Dr John Snow. From his investigation, and the information on cholera mortality provided by William Farr of the General Registry Office, John Snow advanced the theory that cholera was contagious, the infection being transmitted by the contamination of river and well water sources of drinking water by the sewage from infected persons. However, it was not until the cholera outbreak of 1892 in Hamburg, resulting in 8000 deaths, that the cholera vibrio was isolated by Koch from Hamburg's water source - the River Elbe. The outcome of Snow's work was that the water supply intakes of London on the River Thames were moved from the polluted tidal reaches to the much less polluted reaches upstream of Teddington Weir, and sand-filtration of the abstracted water was initiated.

Less spectacularly, but still deadly, the waterborne transmission of enteric fever (typhoid and paratyphoid) exerted its toll in London and other cities in Europe and America. The action taken in London, and extended to other cities in England to mitigate cholera outbreaks, also diminished the threat of enteric fever. Table 1 shows the decline of mortality due to enteric fever in England and in the USA[1].

Although these efforts to control the pathogenic safety of public water supplies were effective in eventually eliminating cholera in the UK - the last outbreaks were in 1893, involving 287 cases and 135 deaths - the complete control of enteric fever was not achieved.

Table 1 Mortality from Enteric Fever, UK and USA (per 10^6 population, all causes 1871-1955)

Year	UK (Typhoid and Paratyphoid)	USA* (Typhoid only)
Annual average over years 1871-75	371	Not available
1914	47	N/A
Annual average over years 1910-14	N/A	150.4
Annual average over years 1934-38	4.6	9.4
1945	<1	2.2
1955	0.3	0.3

* USA figures relate to 78 cities

The first practical installation of chlorine (as bleaching powder) in the emergency disinfection of public water supply was at Maidstone in 1897. A solution of sodium hypochlorite was used in a typhoid epidemic in Lincoln in 1905. In 1910 the first full-scale plant for the chlorination of public water supply was installed at Reading, and rapid development of chlorination of river-drawn water supplies followed throughout the country. However, it was many years before effective chlorination of remote upland sources of water supply and of deep ground waters was established.

Table 2[2] gives an indication of the incidence of waterborne disease outbreaks in more recent times. Notable among these was the Croydon outbreak of typhoid fever in 1937. This was a large outbreak, involving 341 cases of infection and 43 deaths. The outbreak was the subject of a public inquiry which concluded that a well in the chalk near Croydon had been contaminated by an active carrier of the disease who along with others had been working in the well at the time. While this work proceeded, water from the well, unfiltered and unchlorinated, was being pumped into supply. Arising from this occurrence a review was made of outbreaks of disease transmitted in drinking water between 1911 and 1937. This showed that 20 outbreaks had occurred which involved public supplies, and that 16 of the outbreaks were of typhoid or paratyphoid. Following the Croydon outbreak the disinfection of public water supplies using chlorine greatly increased and is now a standard process. The last recorded outbreak of typhoid involving public supply was in 1959 and of paratyphoid in 1970.

Over the years there has been steady advance in the efficiency, effectiveness, and reliability of water chlorination technology. Except for the smallest installations, the original method of marginal chlorination, namely giving a steady dosage equivalent to 0.1 to 0.2 mg l^{-1} in the water, has given way to the controlled dosage of excess chlorine into the water, followed by at least 30 minutes contact storage and then dechlorination with sulphur dioxide to give a residual free chlorine of about 0.1 to 0.2 mg l^{-1}. Often sufficient ammonia is added to the water to convert the residual chlorine to chloramine before delivery to service reservoirs and distribution. This has the advantages of reducing chlorinous odours at the tap and prolonging the retention of a residual disinfection capacity in distribution mains. The retention of disinfection capacity throughout the distribution system is extremely important in guarding against the inevitable accidents and mistakes which occur from

Table 2 Waterborne Disease from Public Supply, UK 1937-86

Disease	Number of Outbreaks		Number of Cases
	Source Contamination*	Distribution Contamination	
Typhoid Fever			
Croydon	1		341
Other outbreaks		2	72
Paratyphoid Fever	1		90
Bacterial Dysentry	3	1	>5000
Gastro-enteritis	4	1	>3500

* All associated with no or faulty chlorination

time to time and result in contamination of the mains water. However, the use of chloramination has to be carefully controlled, otherwise the process can lead to the formation of nitrite in the distribution system, sufficient to cause the water delivered to the customer to fail the EC standards of drinking water quality. Likewise the addition of excess chlorine to waters which contain substantial concentration of organic matter can be expected to result in the formation of halomethanes to such an extent as to result in failure to meet the EC standards.

For these and other reasons the use of chlorine in water disinfection has in recent years been subject to critical review, but it is unlikely to become redundant in the foreseeable future. It is likely that its use will be increasingly augmented by the use of ozone, or of chlorine dioxide (developed in the USA from 1944 onwards). Without any doubt the use of chlorine in water disinfection over the last century has been one of the greatest advances in water purification and has made an immense contribution to environmental health. An important milestone indeed for the chemical industry.

The use of ozone in the treatment of public water supplies began at around the same time as the use of chlorine. The first installation was in Holland in 1893, although an installation built at Nice in 1906 is still using ozone[3]. Over 1000 treatment plants now exist, mostly in France, Switzerland and Germany. There are very few installations in the UK, mainly because an installation of ozonation plus chlorination at Loch Turret in Scotland produced major problems, not of disinfection but of other consequences of ozonation of highly coloured waters. Used correctly, with appropriate post-ozonation filtration, ozone at a concentration of 3 to 5 mg l^{-1} in water with a contact time of 5 to 10 minutes is a good purification process.

Unfortunately, ozone in water is short lived, and to maintain disinfection in distribution systems it is necessary to give a final dose of chlorine, chloramine or chlorine dioxide. In addition to its disinfection value, ozonation is most useful in destroying the stable organic colour of waters derived from either natural sources or from pollution. This results in much of the organic matter being transformed into simpler biodegradable substances which, if not removed by slow-rate filtration before the water passes into distribution, will foul the distribution system and stimulate the growth of the fauna of that system to infestation levels. Ozonation also precipitates iron and manganese from the water which also has to be removed before passage into supply. It was these phenomena, uncontrolled because no filtration was provided, which were the main cause of the Loch Turret problems referred to above.

In general it can be said that in the case of the smallish waterworks operating with a source water of high quality and with a small, sound distribution system, ozonation is likely to produce a better overall product than chlorination. For the best results at all other works (the vast majority), after preliminary removal of particulate matter from the water, ozonation followed by slow-rate filtration through sand or granular active carbon, and a final low-dose chlorination would be the preferred option. Thus it is apparent that while the ozone disinfection of water is not currently in vogue in the UK, there is a strong prospect that its future use worldwide as a treatment process will become a milestone of progress in applied water chemistry.

Other Chemical Processes of Purification of Public Water Supply

The use of chemical coagulation in its various forms for the clarification of public water supply is a worldwide practice. However, this process is not the only one available to give good clarification of water; slow sand-filtration is a sound and a more robust process of clarification but requires much more space and is rather labour intensive. The softening of public water supply using excess lime, lime-soda, or ion exchange is no longer in fashion worldwide. It is not necessary in environmental health terms, and the availability of small, efficient and relatively inexpensive domestic ion exchange units to meet individual needs makes waterworks softening redundant. Treatment plant is available for the reduction of nitrate concentrations in public supplies to meet EC standards. Where this is necessary it is likely to be a stop-gap arrangement until land use changes are enforced to halt, and eventually reverse, the rapid rise in nitrate levels in waters, particularly groundwaters, which has occurred over the last half-century.

The fluoridation of public water supplies to reduce dental caries remains a controversial subject in which it seems that opposition in principle to the mass medication involved is the main obstacle to progress. Research began almost a century ago in the USA into the role of fluoride in water in causing the mottling of teeth. By the mid-1930s it was established[4] that more than 1.5 mg l^{-1} of fluoride in drinking water gave rise to mottling, the severity of the mottling increasing with concentration. Further that children with mottled teeth suffered a lower incidence of dental caries than the average. By 1945 it was established that the addition of fluoride to water to give a concentration of 1 mg l^{-1} would be of great benefit to children's teeth. Three trial studies of the effect of fluoride addition to water supplies were then set up in the USA, the most famous perhaps being at Grand Rapids, Michigan where the results of 11 years' study were reported in 1957. These results showed that from yearly dental examinations of children aged between 4 and 16, a 60% or more reduction in dental caries had occurred.

The first trials of the effect of fluoride addition to public supply in the UK were set up at Watford, Anglesey and Kilmarnock in 1955/56, and the results obtained were in line with the USA findings. As regards possible adverse effects from fluoridation of public water supplies, many detailed medical investigations have been carried out and there is no evidence that the presence of 1 mg l^{-1} of fluoride in public water supply has any adverse effect on health. Nevertheless, the recommendations of the WHO[5] on drinking water quality put a limit of 1.5 mg l^{-1} on fluoride and indicate that at fluoride concentrations of 3 mg l^{-1} damage to health may arise.

Progress in the application of fluoridation in public supply in the UK has been very slow. Much greater progress has been made in the USA and many other countries. This is clearly a case where a new milestone of the progress of chemistry in improving environmental health has been hewn, but this milestone cannot be erected in the UK until the people require this to be done.

3 CHEMISTRY IN WATER POLLUTION CONTROL

Chemistry has played a fundamental role in development of the control of water pollution. This development was formally commenced on a national basis by the enactment of the Rivers Pollution Act 1876, although the earliest statute on river pollution prevention was enacted during the reign of Richard II (1388). This prohibited the dumping of refuse into ditches and rivers near to towns. The 1876 Act prohibited pollution caused by the dumping of refuse or by discharge or deposit of sewage matter, industrial waste and mining waste. By and large this Act was a sound legal measure, but it had two characteristics which greatly diminished the effectiveness of its enforcement. The first of these shortcomings, which could hardly be avoided having regard to the embryo state of waste disposal technology of the times, can now be seen to have been the grandfather of BATNEEC (best available technology not entailing excessive cost), that albatross currently hanging around the neck of Her Majesty's Inspectorate of Pollution. This arose inasmuch as the 1876 Act made provision whereby river pollution could not be penalised if the offender could show that he had used the best practicable and available means to render the polluting matter harmless. The second and major shortcoming of the 1876 Act was that its enforcement was placed in the hands of the local sanitary authorities, who were themselves responsible for sewage disposal and hence for much of the pollution occurring (which error was repeated between 1974 and 1989 in the era of integrated water management of the Water Authorities, now defunct). The current law relating to water pollution control is embodied mainly in the Water Act 1989.

In the great and fascinating saga of the role of chemistry in water pollution, there are two milestones of progress in which the UK led the world. The first relates to the treatment of the biodegradable organic content of wastewaters using the Activated Sludge Process, and the second relates to the clean-up of the River Thames in London.

The Activated Sludge Process

The basis of this process was established in 1914 by Ardern and Lockett[6] at Manchester, and developed in the nineteen twenties by the Activated Sludge Company and others. Essentially the process consists of aerating a continuous flow of sewage or other degradable waste waters to produce a sufficient concentration of sludge 'floc' of bacteria, fungi, and protozoa (activated sludge) which removes most of the degradable organic and nitrogenous matter from the wastewater. The continuous production of activated sludge is removed from the treated liquor by gravity settlement, a proportion of the separated sludge is recycled to treat more wastewater and the remainder is given further treatment before disposal. The sludge-free treated wastewater is usually sufficiently purified to be discharged direct to a river or sea outfall.

The first large installation of this process was at the Twickenham (Mogden) sewage works of the Middlesex County Council in 1936[6] and remains in operation today treating some 0.6×10^6 m^3 d^{-1} of mixed sewage and industrial wastewater. Following the success of this installation, the process was developed worldwide to become almost universally used in the larger conurbations. Many variations of the process have been developed for particular purposes, including use in small scale applications. Without doubt the original and development work of the chemists, and chemical and civil engineers involved produced a major step forward, which now awaits transformation through application in the future of the results of molecular biochemistry and genetic engineering.

The Clean-up of the Tidal River Thames

The saga of the pollution history of the River Thames is both extensive and well recounted. The gross pollution of the Thames tideway reached its peak in 1858, the year of 'The Great Stink', when a hot summer and parliamentary and public outcry forced the taking of purposeful remedial action. The great scheme of Sir Joseph Bazalagette, the Chief Engineer of the then Metropolitan Board of Works, was put in hand. This was constructed to convey the drainage of London eastwards in new intercepting sewers along the north and south banks of the tideway, to discharge near Barking and Woolwich respectively. On completion of this work, the quality of the tideway through London was much improved but the position steadily worsened until the mid-1960s when decisions were taken to provide for full purification of London's sewage using the activated sludge process referred to above. By 1979 some £1 billion (at current prices) had been spent in achieving a massive improvement of the entire tideway, and the first substantial restocking of the freshwater river with juvenile salmon was begun.

Crucial to these results was the work of the former Water Pollution Research Laboratory[7], which in 1964/65, for the first time properly explained the chemistry of tidal water pollution, and indeed of nitrification and de-nitrification in water generally, and produced the first computerised predictive model of a polluted estuary. The ecological proof of successful application of the results of this research came in July 1982 when the first ascent of adult salmon into the freshwater river for 150 years was observed at Hampton Court (Molesey) Weir. Many chemists, engineers and biologists can take great pride in their contribution to this milestone of progress in environmental health.

REFERENCES

1. 'Water Treatment and Examination', W.S. Holden, Churchill, London 1970.

2. N.S. Galbraith, N.J. Barrett, R. Stanwell-Smith, J.Inst. Water and Env. Man., 1987, 1, 7.

3. R.A. Hyde, T.F. Zabel, 'Uses of Ozone in Water Treatment', Seminar on Ozone in UK Water Treatment Practice, Inst. Water Engineers and Scientists, London 1984.

4. H.T. Dean, Pub. Health Reports, 1938, 53, 1443.

5. 'Guidelines for Drinking Water Quality', WHO, Geneva, 1984, Vol.1, 55.

6. E. Ardern, W.T. Lockett, 'Experiments on the Oxidation of Sewage Without the Aid of Filters', J.Soc.Chem.Ind. 1914, 33, No.10.

7. 'Effect of Polluting Discharges on the Thames Estuary', Notes on Water Pollution Research Technical Paper No.11, HMSO, 1964.

The Quest for New Medicines

B.J. Price and M.G. Dodds

GLAXO GROUP RESEARCH LTD., GREENFORD, MIDDX. UB6 0HE, UK

1 HISTORICAL PERSPECTIVE 1841-1991

150 years ago the Apothecary's shop was filled with staple remedies including ipecac, morphia, opium, quinine, squill, cinchonidia sulphate, cloves, ergot and belladonna. Contemporary advances in medicine involved the discovery of the anaesthetic properties of nitrous oxide (Wells,1845), ether (Morton,1846) and chloroform (Simpson,1847), agents which heralded the way for major technical advances in surgery and freedom from pain for the patient.

Now, when the Society is celebrating its 150th Anniversary, patented medicines for a £77 billion world market are manufactured by multinational corporations in a regulated industry with a large R & D sector. Although drugs are now available in every therapeutic area and have contributed importantly in many countries to reduced infant mortality and increased life expectancy, much remains to be achieved. Contemporary medical advances include organ transplantation and the discovery of viral oncogenes; new medicines entering into use include recombinant proteins (insulin, interferons, interleukin-2, erythropoietin).

Four Phases in the Growth of Pharmaceutics

The First Phase in the growth of pharmaceutics involved the long slow accumulation of knowledge and use of folk medicines, many of them derived from plants, the production of which had become increasingly standardised in pharmacopoeias by the middle of the 19th Century.

The Second Phase related to the increasing importance of pure natural products, which were isolated from plants (e.g. the alkaloids colchicine, emetine) and animals (e.g. heparin, insulin, sex hormones, vitamins). Penicillin, isolated by Fleming in 1929, and characterised and developed by the Oxford group of Florey and Chain more than 10 years later, was the first of numerous antibiotics to be obtained from micro-organisms.

The Third Phase saw the period of rapid growth of synthetic medicinal chemistry. It was ushered in by the pioneering works of Ehrlich (salvarsan), Dreser (aspirin) and Fischer (barbitone) at the beginning of this century, and continues through to the present. Some of the 'milestone' compounds in this phase came from within the medicinal chemistry industry itself (e.g. organomercurial diuretics, antihistamines, benzodiazepines). Others were consequent on basic science discoveries. Thus knowledge about the nature of neurohumoral transmission and the transmitter substances (from the work of Barger & Dale, Loewi, Cannon, von Euler) eventually led to the discovery of new therapeutic substances acting at cholinergic, adrenergic and other receptors.

During this period, successive medicinal chemistry discoveries contributed importantly to the treatment of a wide range of infectious diseases. Salvarsan (for syphilis), suramin (trypanosomiasis), quinacrine (malaria), prontosil and the sulphonamides (antibacterial chemotherapy), thiabendazole (anthelmintic), dapsone (leprosy) and isoniazid (tuberculosis) are some of the celebrated discoveries that have done so much to relieve suffering. These drugs have been of enormous social and economic importance.

The Fourth Phase of growth in pharmaceutics is the most recent, during which time medicinal chemistry has built upon the advances in molecular biology and cell chemistry. Stepping-stones in basic science during this period have included an understanding of DNA structure, enzyme active sites, the nature of adaptive immunity, the mechanisms of viral replication, the discovery of oncogenes and cytokines etc. Already a whole new range of compounds have appeared including enzyme inhibitors (allopurinol, captopril), antiviral (amantidine, acyclovir) and anti-neoplastic agents (cyclophosphamide, adriamycin, cisplatin), and many more compounds will emerge shortly targeted to diseases of the immune system.

2 MILESTONES IN MEDICINAL CHEMISTRY

Drugs Acting at Receptors

The accumulation of knowledge concerning the mechanisms of neurohumoral transmission and the identity of the transmitter substances led to advances in medicinal chemistry which have provided a whole range of compounds acting at receptors. Some of the major developments concerning adrenoceptors, histamine and 5-HT receptors are described below. Each arose from collaboration between the chemists who produced the molecular probes and the pharmacologists who determined their activities.

In 1905, Langley had conceived of two types of tissue 'receptors', excitatory and inhibitory, to explain the actions of catecholamines, and Dale's demonstration in 1927 that ergot alkaloids blocked the excitatory actions of adrenaline (e.g. constriction of blood vessels) but not the inhibitory actions (e.g. relaxation of bronchial smooth muscle) provided the evidence for Langley's hypothesis. The next landmark event was in 1948 when Ahlquist proposed an adrenoceptor classification comprised of two receptor types α and β based on the tissue responses to adrenaline (1), noradrenaline and isoprenaline. However, his observations were neglected for 10 years until the compound dichloroisoprenaline (5) was reported (Powell & Slater, 1958; Moran & Perkins, 1958). The activity of this β-adrenoceptor blocking compound (a partial agonist and a pharmacological tool rather than a therapeutic agent) supported Ahlquist's hypothesis. The classification of β-adrenoceptors into β_1 and β_2 subtypes by Lands et al. (1967) provided the explanation for the bronchial smooth muscle (β_2- mediated) specific action of salbutamol (3) and similar compounds.

β_2-Adrenoceptor Agonists. The first β- adrenoceptor agonists used as bronchodilators in treating asthma (isoprenaline (2) and orciprenaline) were short-acting and their use was associated with cardiac stimulation and skeletal muscle tremor. During 1966, in their search for compounds more selective for bronchial muscle, Brittain, Jack, Lunts and their colleagues found that the saligenin analogue of isoprenaline, was more active on isolated guinea pig trachea than atrial muscle, and that the N-t-butyl analogue, salbutamol (3), was even more selective. Inhalation therapy with salbutamol and terbutaline (4) enables relief of symptoms to be achieved for useful periods without patients suffering cardiac stimulation or skeletal muscle tremor.

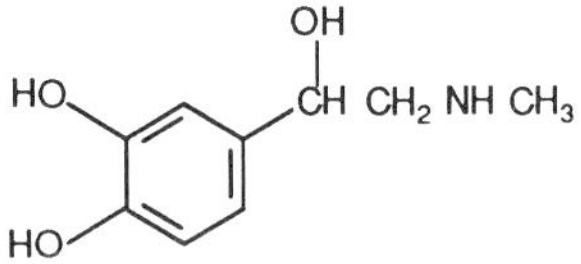

OH, HO, HO, $CH\ CH_2\ NH\ CH_3$

(1) Adrenaline

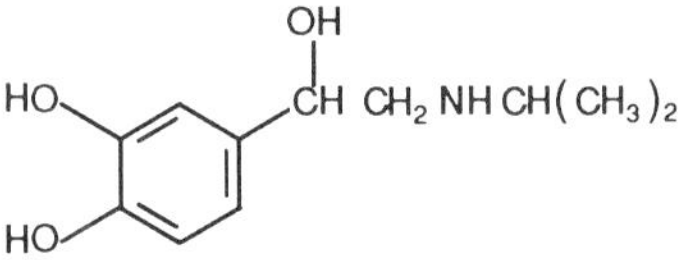

OH, HO, HO, $CH\ CH_2\ NH\ CH(CH_3)_2$

(2) Isoprenaline

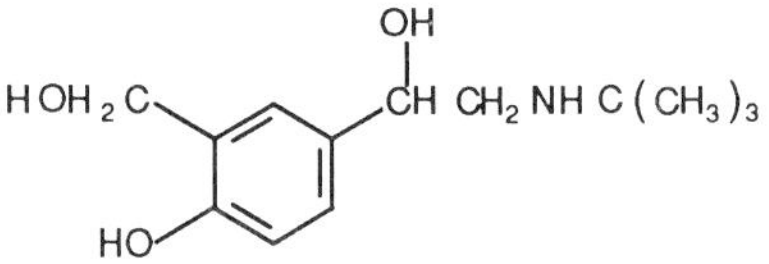

OH, HOH_2C, HO, $CH\ CH_2\ NH\ C(CH_3)_3$

(3) Salbutamol

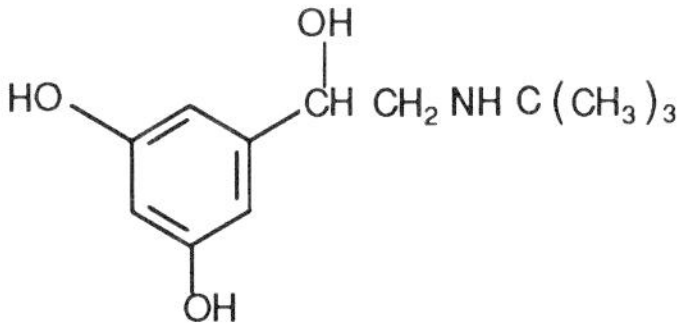

OH, HO, OH, $CH\ CH_2\ NH\ C(CH_3)_3$

(4) Terbutaline

OH, Cl, Cl, $CH\ CH_2\ NH\ CH(CH_3)_2$

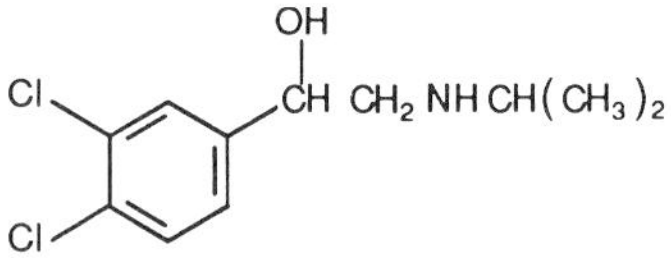

(5) Dichloroisoprenaline

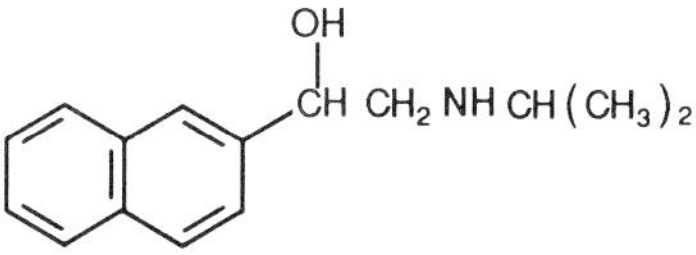

OH, $CH\ CH_2\ NH\ CH(CH_3)_2$

(6) Pronethalol

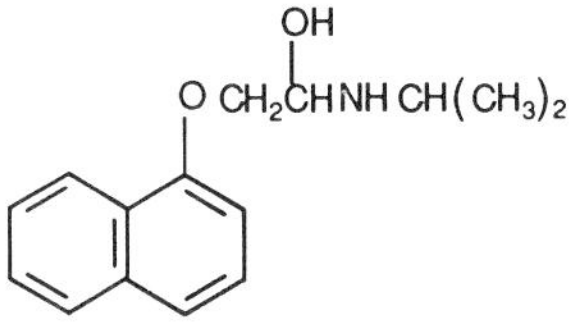

OH, $O\ CH_2CHNH\ CH(CH_3)_2$

(7) Propranolol

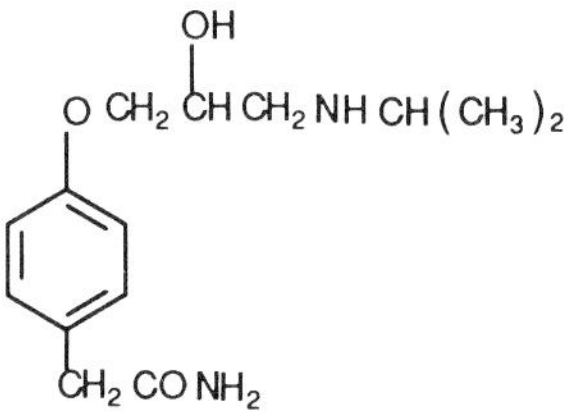

OH, $O\ CH_2\ CH\ CH_2\ NH\ CH(CH_3)_2$, $CH_2\ CO\ NH_2$

(8) Atenolol

β-Adrenoceptor Antagonists. After the discovery of dichloroisoprenaline (5), Black (now Sir James), working with the chemists Stephenson and Crowther at ICI, had the vision to see the therapeutic potential of blockade of cardiac excitatory responses. The team succeeded in bringing forward first pronethalol (6), which proved to be carcinogenic in mice, and then propranolol (7), which was marketed in 1964 for the treatment of angina pectoris. Subsequently, propranolol became important in the treatment of hypertension, following the clinical studies of Pritchard & Gillam (1964).

Many thousands of compounds have been synthesised and numerous blockers have come onto the market over the last 25 years. Major issues have included the advantage or otherwise of intrinsic sympathomimetic (partial agonist) activity, CNS penetration and cardioselectivity. Perhaps the most important compounds to follow the 'non-selective' β_1,β_2 antagonist propranolol were: practolol, cardioselective β_1-adrenoceptor antagonist with weak partial agonist activity (the first to overcome the problem of non-selective β-blockers precipitating asthma in susceptible individuals as a consequence of β_2-blockade); atenolol (8) and metoprolol, cardioselective with no agonist properties (widely used in the treatment of hypertension).

Histamine H_1 Antagonists. Dale had isolated histamine (9) from liver and lung tissue extracts in 1927, and Lewis had demonstrated the release of 'H' substance from cells in the skin following injury at about the same time. In 1937 Bovet & Staub described the first compounds with antagonist activity, which protected guinea pigs against anaphylactic shock and antagonised the constrictor effects of histamine in various types of smooth muscle. Numerous compounds were introduced subsequently for the treatment of allergy and motion sickness. Mepyramine (10) is a typical example of the class of compounds which specifically antagonise the H_1 receptors subsequently identified by Ash & Schild in 1966.

Histamine H_2 Antagonists. In 1964, Black had moved to Smith, Kline & French and begun to look for treatments for peptic ulcer disease. Histamine was a potent stimulant of gastric acid secretion and with the knowledge that 4-methyl histamine behaved as a selective agonist at these novel H_2 histamine receptors, Black and his team pursued the search of several hundred compounds synthesised by Ganellin and his chemists. Using bioassay, in particular the ability of compounds to inhibit histamine-induced gastric acid secretion in anaesthetised rats, they eventually were successful. The discovery of the partial agonist activity of 9α-guanyl histamine, proved to be the harbinger of eventual success, since it led to the discovery of full competitive antagonists of the histamine H_2 receptor. The first compounds, burimamide and metiamide, were not marketed (lacking sufficient oral activity in the case of burimamide, and because of granulocytopenia in the case of metiamide). However cimetidine (11) proved a breakthrough in the treatment of ulcer disease. It became the first $1 billion product and first it and then ranitidine (12), which had some advantages, became the best-selling prescription medicines in the world. In the case of ranitidine the heterocycle is a furan ring, unlike the early series of H_2 blockers which were imidazole, pyridine or thiazole derivatives.

At present selective agonists and antagonists for the H_1 and H_2 subtypes are known and there continues to be speculation about a third subtype, for which (R) α-methyl histamine may be a selective agonist. This H_3 receptor may be present as an autoreceptor in histaminergic nerve terminals in the CNS, and thus selective antagonists might be useful in cognitive disorders.

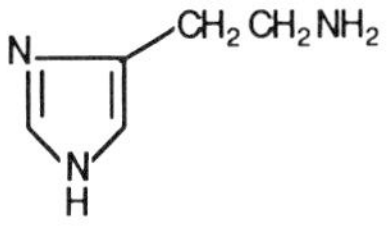

(9) Histamine

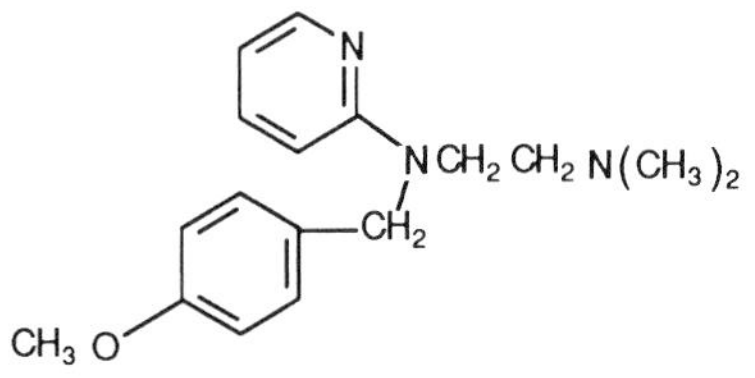

(10) Mepyramine

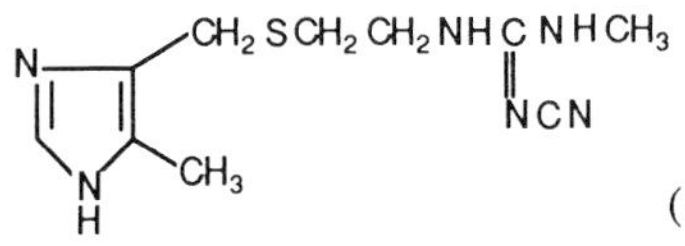

(11) Cimetidine

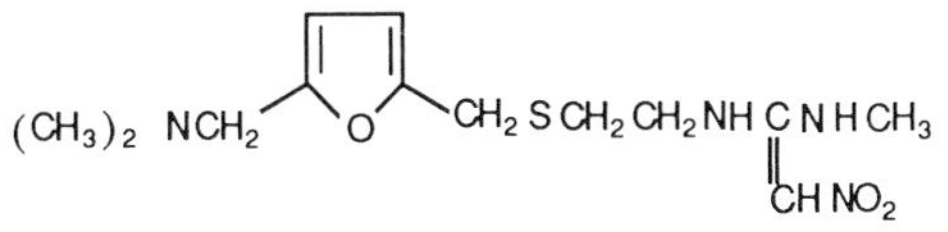

(12) Ranitidine

Other Major Drug Milestones

Several important new medicines came from empirical searching rather than from rational programmes of drug discovery. Some emerged from within industry, others stemmed from clinical observation and research; some came from the work of a distinguished pioneer, others were corporate rather than individual achievements. Aspirin, the thiazide diuretics, the psychoactive and neuroleptic drugs, the anti-inflammatory steroids and the oral contraceptives all came forward in the Third Phase of Growth. All were significant therapeutic advances but their origins and course of development were very different.

The Non-steroidal Anti-inflammatory Drugs (NSAIDs). Salicin, the glycoside of salicylic acid and active principle of the extract of willow bark that Edmund Stone had reported to reduce pain and fever in 1757, was isolated by Buchner in 1828. Aspirin, acetylsalicylic acid (13), was discovered in 1898 by Hofmann. It was first manufactured by Bayer, one of the great German enterprises that grew out from the aniline dye industry. The antipyretic, analgesic and anti-inflammatory activities of aspirin led to its becoming the most widely used and durable household remedy (with 16,000 tons of tablets consumed in a year in the US alone). Other aspirin-like drugs were developed in large numbers, including carboxylic acid derivatives (e.g. indomethacin (14), ibuprofen (15), the fenemates), pyrazoles (e.g. phenylbutazone) and oxicams (e.g. piroxicam). The mechanisms responsible for the biological activities of the NSAIDs remained a mystery until 1971, when Vane (now Sir John) and his colleagues provided evidence that they behaved as inhibitors of the cyclo-oxygenase enzyme which converts arachidonic acid to the cyclic endoperoxide prostaglandins PGG_2 and PGH_2. Blockade of prostaglandin synthesis and release following tissue injury appeared to explain all the biological activities of the NSAID drugs and their side effects. However, more recent evidence suggests that at least some of their activities may be related to their ability to inhibit neutrophil adhesion and activation. The existing NSAID drugs are limited to symptom relief rather than disease control, so the challenge remains to obtain non-steroidal compounds with efficacy at least comparable to that of the corticosteroids.

Diuretics. The thiazide diuretics have occupied an important position in the management of oedematous states and hypertension ever since their discovery by Karl Beyer and his colleagues in 1958. The synthesis by the Merck chemists of compounds related to the carbonic anhydrase inhibitor acetazolamide, led to the observation that some benzothiadiazides produced diuresis and natriuresis independent of enzyme inhibitory action. These diuretics, analogues of 1,2,4-benzothiadiazine such as chlorothiazide (16), increased Na^+ and Cl^- ion excretion through a direct action on distal renal tubular sodium reabsorption. They became the standard first treatment step in the management of hypertension, following evidence of efficacy and reduced cardiovascular complications in clinical trials organised by the Veterans Administration. Later, other compounds were discovered with greater diuretic efficacy, related to an action on the

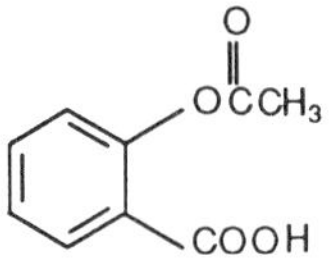

(13) Acetylsalicylic acid

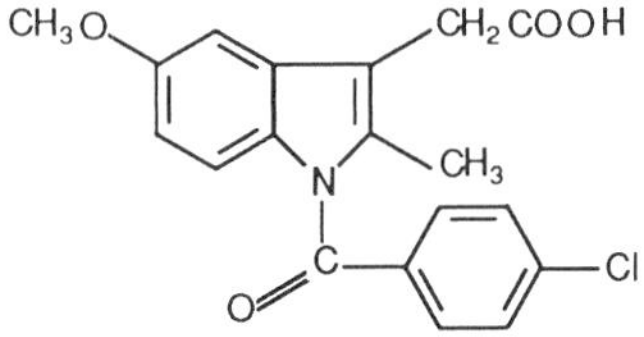

(14) Indomethacin

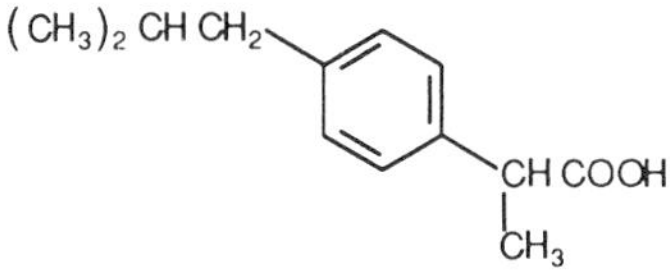

(15) Ibuprofen

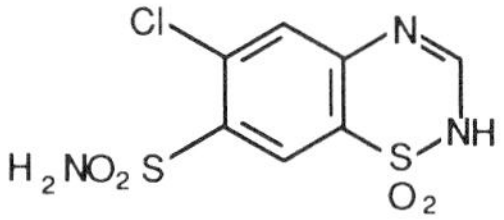

(16) Chlorothiazide

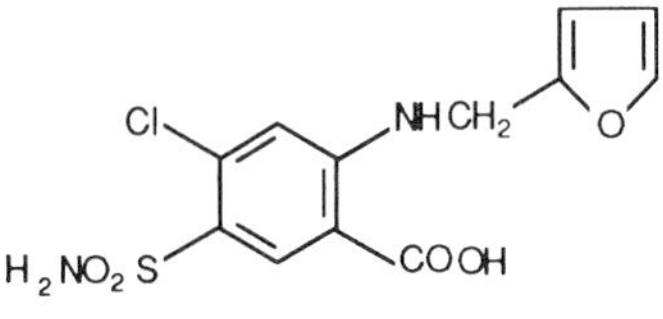

(17) Frusemide

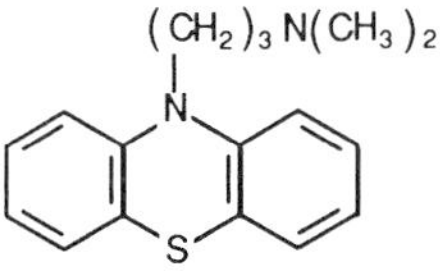

(18) Chlorpromazine

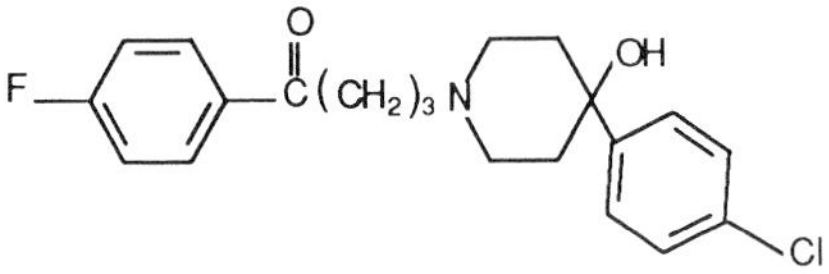

(19) Haloperidol

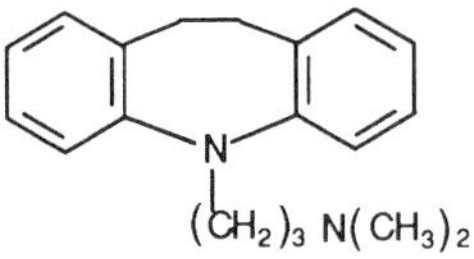

(20) Imipramine

ascending limb of the loop of Henle. The most important, frusemide (17), is of particular value in acute pulmonary oedema.

Psychotropic and Neuroleptic Drugs. Modern psychopharmacology began with the synthesis of chlorpromazine (18) by Charpentier in 1950. It ameliorated various psychoses and became the prototype for numerous compounds used in psychiatric disorders. One of the more important compounds to emerge was the neuroleptic haloperidol (19). This butyrephenone analogue was discovered by the Belgian group at Janssen and its antipsychotic effects were first demonstrated in 1958. Several drug leads in the CNS area also emerged from serendipitous discovery of benefit in the course of clinical treatment of other diseases. Thus the elevation in mood observed in tuberculous patients treated with iproniazid, led to the development of antidepressant monoamine oxidase inhibitors such as phenelzine and tranylcypromine. In similar manner, the antidepressant activity of imipramine (20) was discovered in 1958 in the course of a clinical trial. Unlike the phenothiazines, this compound (in which the sulphur is replaced by an ethylene link in the central ring), was ineffective in agitated psychotic patients but improved the condition of certain depressed patients.

Anxiety might be regarded as a natural part of the human condition but relief sought through the 'tranquillising' benzodiazepines led to their becoming some of the most commonly prescribed medicines in the 1960s and 70s. The first successful compound, chlordiazepoxide (21), was developed by Sternbach's group at Hoffmann-La Roche in the late 1950s. Numerous related compounds were synthesised, of which diazepam was the most important. The knowledge that habituation can develop and that withdrawal symptoms can occur on stopping treatment, sometimes with rebound anxiety, contributed to a decline in their use. Over the last 20 years there have been fewer advances in CNS drug therapy. With increasing knowledge of the transmitters involved in mental illnesses and availability of receptor subtype-specific ligands (for dopamine, 5-HT, histamine, acetylcholine, neurokinin and excitatory amino acid receptors) some novel treatments might be anticipated in the next decade.

Cortisone Synthesis. Sarett (1946) achieved the first synthesis of cortisone (22), but most of the reactions proceeded with poor yields and the method was unsuitable for production. In March of the same year, Merck undertook the laboratory production of cortisone starting from desoxycholic acid. The first material came through in April 1948. In May, Hench and Kendall started clinical studies at the Mayo Clinic and published their results in April 1949. The Fiesers have told how Merck then took the unprecedented step of adapting to large scale production a multi-stage process which used very expensive reagents requiring critical control. The work under Tishler succeeded however and cortisone acetate was made available to physicians by the end of 1949 at $200/gram. By improvements to the process and with competition from other manufacturers, the price had fallen to $2/gram in less than 10 years and the sales volume of corticosteroid drugs had risen to $100 million a year.

Alternative possible routes to cortisone not dependent on bile acid starting materials included introducing an oxygen atom at C11 into abundant sterols or sapogenins such as cholesterol, stigmasterol or diosgenin, or using hecogenin which had a 12-ketone group in ring C, or total synthesis. The route from hecogenin was the only one eventually used on a commercial scale. A useful source of hecogenin was found in the sisal plant Agave sisalana, which was cultivated widely in East Africa. Solvent extraction and acid hydrolysis enabled hecogenin to be obtained in quantity from the whole plant or sisal waste. The conversion of hecogenin to the intermediate 11-ketotigogenin was tried in various ways but the Glaxo group made improvements to the initial bromination step and developed an efficient conversion of the 11,12-ketol 3,21-diacetate to the corresponding 11-ketone. This was converted into cortisone by well developed methods, and the whole process of preparation from hecogenin provided a means of commercial manufacture. Subsequently, an Upjohn group led by Peterson found soil micro-organisms (e.g. Rhizopus nigricans) which hydroxylated progesterone to 11α-hydroxyprogesterone, an early and important example of biotransformation. Syntex later used this as an intermediate in a route to cortisone.

Anti-inflammatory Corticosteroids. Building on the intermediates afforded by the synthesis of cortisone, industry began the search for compounds with increased activity. The first commercially successful anti-inflammatory compounds were the 1-dehydro derivatives, prednisone and prednisolone, obtained by the Schering Corporation. These were succeeded by the even more potent 9α-fluoro compounds triamcinolone, dexamethasone (23) and betamethasone (24). The systemic use of these agents led to side effects such as atrophy of the adrenal cortex and suppression of the immune system that largely precluded their use in conditions such as asthma, eczema and psoriasis. Topical application of these steroids in creams or ointments however relieved skin conditions and more lipophilic compounds (e.g. fluocinolone acetonide) were subsequently developed which accessed the epidermis but did not pass readily into the dermis and systemic circulation. The Glaxo group marketed betamethasone 17-valerate in 1963 and 11-keto 21-halo 17-esters of

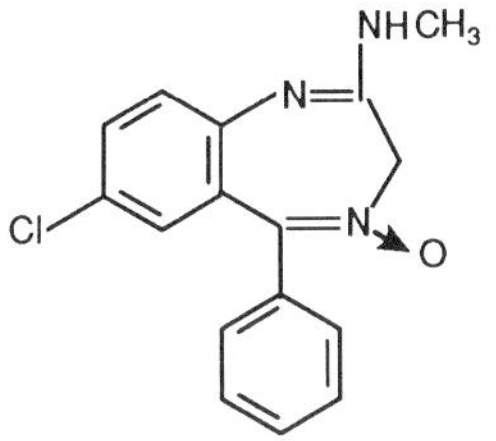

(21) Chlordiazepoxide

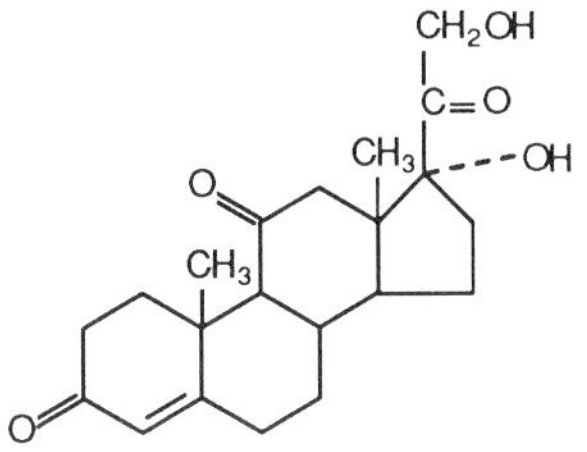

(22) Cortisone

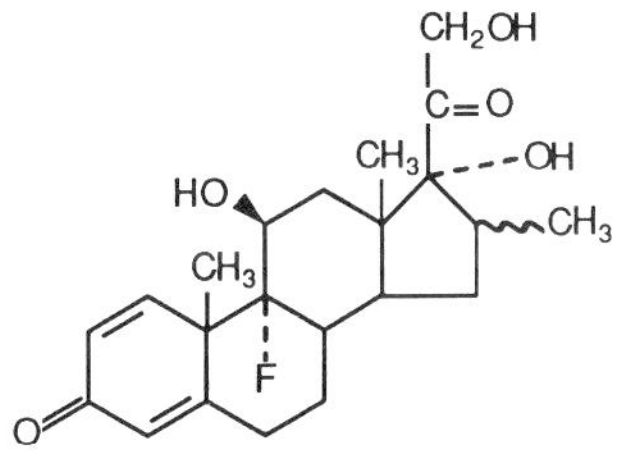

(23) 16α : Dexamethasone

(24) 16 β : Betamethasone

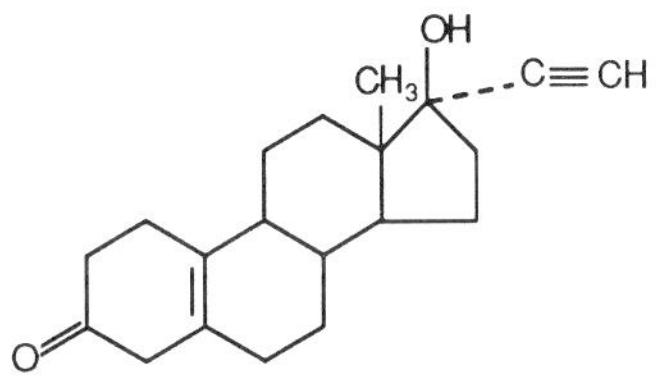

(25) Norethynodrel

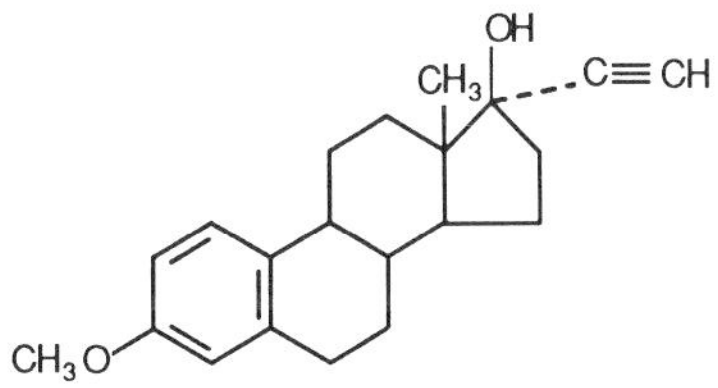

(26) Mestranol

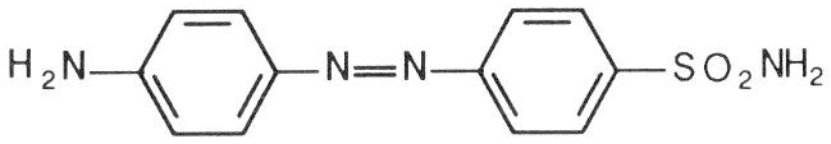

(27) Prontosil

betamethasone and related steroids were developed subsequently with improved activity and minimal side effects. Direct delivery to the lungs, by means of a metered dose inhaler, enabled preparations of some of these steroids (e.g. beclomethasone 17,21-dipropionate) to be used also in the treatment of asthma. Symptoms could be controlled without suppression of adrenal function or other major side effects.

Contraceptive Steroids. The isolation and characterisation of the naturally-occurring steroid hormones in the 1930s enabled experiments to be performed showing that oestrogens and progestogens could inhibit ovulation in experimental animals. The first comprehensive clinical studies to confirm that a progestational agent could inhibit ovulation were carried out by Pincus and his colleagues in Puerto Rico in 1955 using tablets containing progestogen, norethynodrel (25), and a small amount of oestrogen, mestranol (26). The remarkable efficacy of this method of contraception prompted numerous further developments and widespread use (by 65 million women worldwide in 1985). Norethisterone and norethynodrel, synthesised at GD Searle by the group led by Colton, were the first clinically useful progestational agents. Addition of a very small quantity of oestrogen (in a 'combined pill') helped prevent breakthrough bleeding and mestranol became the most important oestrogen used in contraceptive preparations. Both norethynodrel and mestranol are pro-drugs (of norethisterone and ethinyloestradiol respectively) and several other oestrogens and progestogens developed later were also pro-drug forms of these compounds.

Enzyme Inhibitors

Inhibition of enzyme action has provided numerous important therapeutic substances. The sulphonamides (inhibitors of dihydropteroate synthase), penicillins (inhibitors of the terminal stage of peptidoglycan synthesis) and clavulanic acid (inhibitor of β-lactamase) are examples. The inhibitors described below are landmark compounds in the fields of antibacterial and cancer chemotherapy, immunoregulation, viral disease and hypertension.

Sulphonamides. Effective antibacterial chemotherapy began with these drugs. In 1932, Domagk discovered that prontosil (27), an azo dye, protected mice against streptococcal and other infections; in 1936 Colebrook and Kenny and others confirmed its efficacy in human infections. In the tissues the azo group of prontosil is split to yield p-aminobenzenesulphonamide (sulphanilamide), the active moiety, and thousands of derivatives of this compound were synthesised and tested in the next decade. A few of them (e.g. sulphadiazine, sulphadimidine, sulphafurazole) were of immense importance up until the discovery of penicillin, although they only exerted a bacteriostatic effect and acquired bacterial resistance developed rapidly. The mechanism of action of the sulphonamides was elucidated in the 1940s by Woods and others, in one of the first molecular theories of drug action, by showing that sulphonamides compete with p-aminobenzoic acid for the bacterial enzyme responsible for the formation of 7,8-dihydropteroic acid. Simultaneous administration of a sulphonamide with trimethoprim (an inhibitor of dihydrofolate reductase) was later shown to achieve a useful synergistic antibacterial effect, since it produces a sequential block in the microbial synthesis of dihydrofolate.

Penicillins and Cephalosporins. Biosynthesis of both the penam and cephem ring systems begins with the construction of the tripeptide δ-(l-α-aminoadipoyl)-L-cysteinyl-D-valine and the formation of isopenicillin N. Although methods of total synthesis of penicillins (by Sheehan) and cephalosporin C (by Woodward) were devised, all of the penicillin and cephalosporin antibiotics are produced commercially by semi-synthetic methods from starting materials obtained by fermentation. Efficient methods of isolation of 6-aminopenicillanic acid (28) from fermentation broths were first developed by the Beecham group in 1959. This provided medicinal chemists with almost unlimited opportunity for sidechain modification and thereby a means of enlarging the spectrum of antibacterial activity and combating bacterial resistance. Batchelor, Doyle and their colleagues led these developments with their syntheses of the anti-pseudomonal compound carbenicillin and the orally active broad spectrum antibiotic ampicillin (29).

The subsequent discovery of commercial methods of manufacture of 7-aminocephalosporanic acid (30) from cephalosporin C (using nitrosyl chloride or phosphorus pentachloride) provided a starting material for the synthesis of cephalosporins with modifications at the 3- or 7-positions. The first compounds used clinically, with broad spectrum of activity and resistance to staphylococcal penicillinase, were cephalothin and cephaloridine (31). Cephalexin (32), the first orally-active cephalosporin, was developed in the late 1960s by Glaxo and Eli Lilly. The search for compounds effective against Gram -ve bacteria which produced β-lactamases, eventually led to the development of 'second generation' cephalosporins. Cephamycin C, a natural product obtained from Streptomyces clavuligerus, was a breakpoint in the progress toward compounds with intrinsic resistance to β-lactamases. Modifications to the 7α-amino substituent of cephamycin C led to compounds with improved activity e.g. cefoxitin (33). Cephalosporins with a 2-aryl 2-oximinoacetamido sidechain were discovered

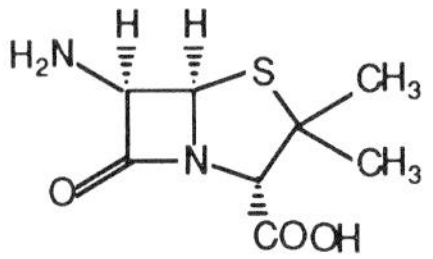

(28) 6-Aminopenicillanic acid

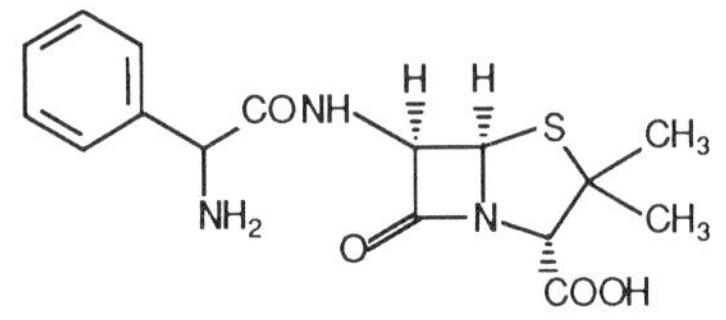

(29) Ampicillin

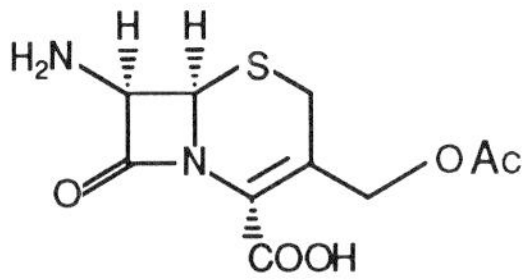

(30) 7-Aminocephalosporanic acid

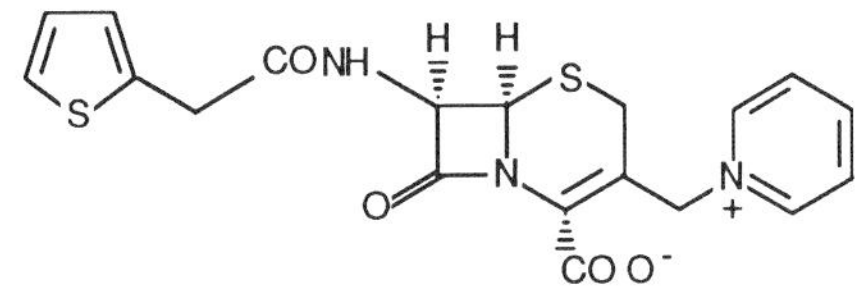

(31) Cephaloridine

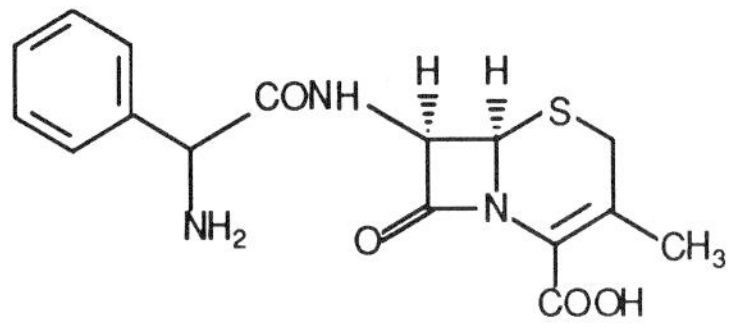

(32) Cephalexin

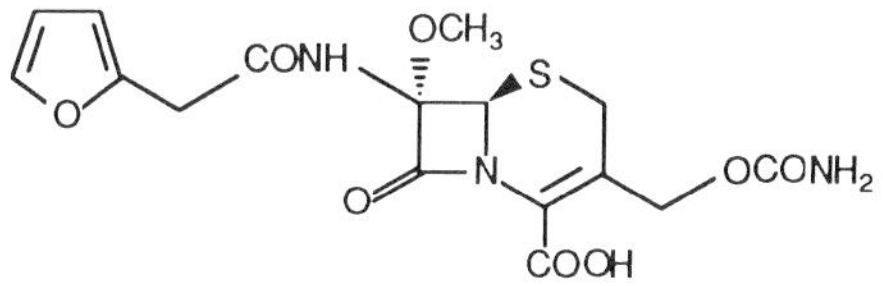

(33) Cefoxitin

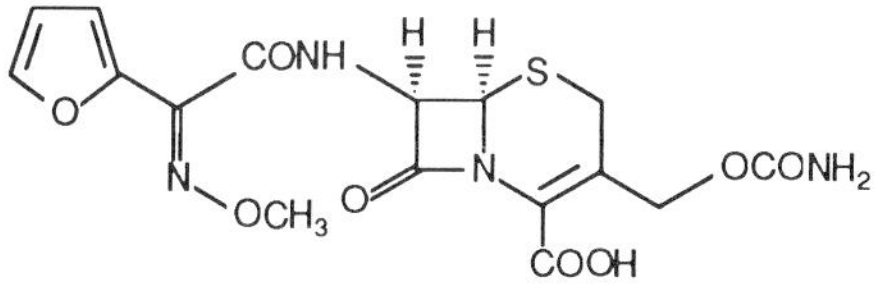

(34) Cefuroxime

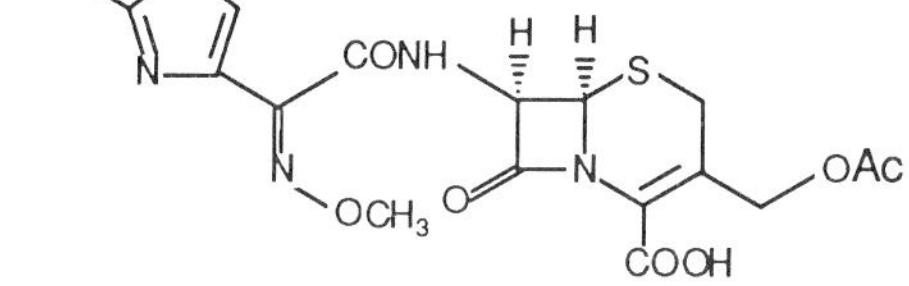

(35) Cefotaxime

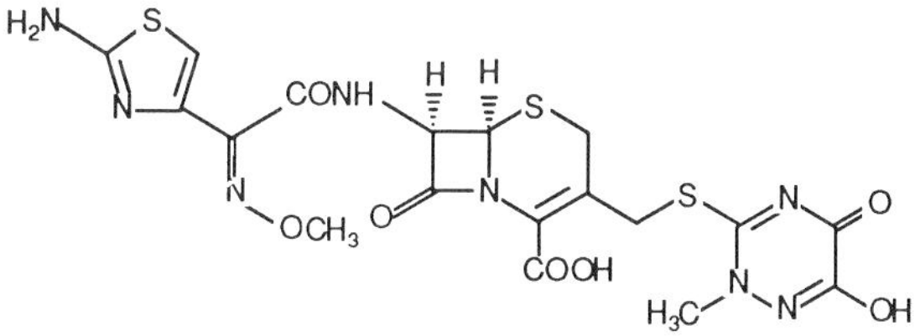

(36) Ceftriaxone

which also exhibited stability to β-lactamases e.g. cefuroxime (34). In them the geometry of the oximino group is important, the *syn*-oximes usually being much more active than the *anti*-isomers. In 1977, Roussel-UCLAF produced cefotaxime (35), the first of a third generation of cephalosporins containing a 2-aminothiazole group in the N-acyl side chain which conferred a substantial improvement in activity particularly against the *Enterobacteriaceae*. The utility of cefotaxime itself was limited by susceptibility of the 3-acetomethoxy substituent to serum esterases. However, numerous other compounds have been synthesised now with better spectrum of activity and improved pharmacokinetics, of which ceftriaxone (36) and ceftazidime (37) are probably the most important examples, the culmination of 30 years of cephalosporin medicinal chemistry.

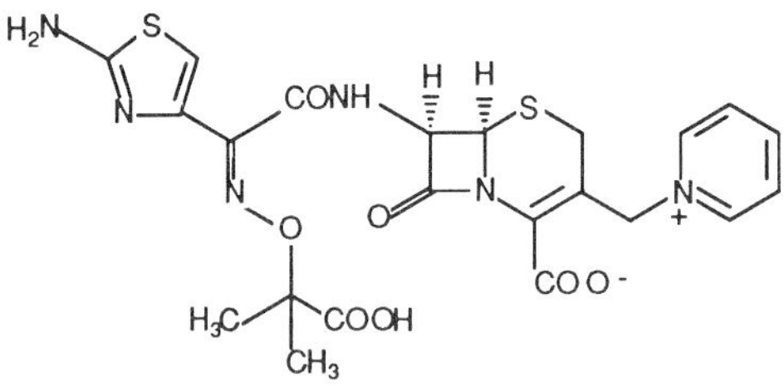

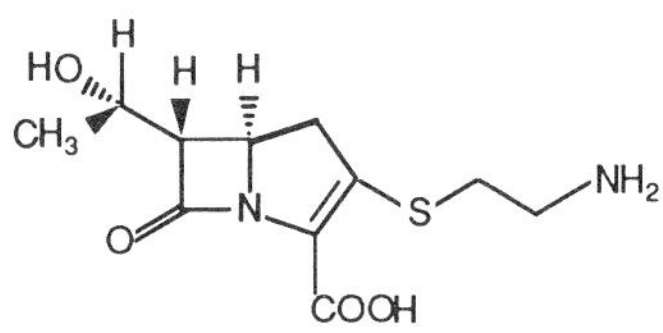

(38) Thienamycin

(37) Ceftazidime

Thienamycin. This broad spectrum β-lactamase-stable antibiotic with anti-pseudomonal activity, was first isolated from *Streptomyces cattleya*. It contained the novel carbapen-2-em ring system (38). However, realisation of its potential as an antibiotic product, and a practical synthetic process for its production, required enormous effort by the Merck group under Christensen.

Chemical instability precluded the commercial development of thienamycin itself (through reaction of the cysteamine side chain with the β-lactam carbonyl). Synthesis of compounds with less nucleophilic basic functions led to the discovery of imipenem (in which the amine was converted to a formimidoyl function), which was stable chemically but with similar or better antibacterial activity. However thienamycin, imipenem and related carbapenem antibiotics were rapidly metabolised *in vivo*, and $<10\%$ of a dose could be recovered from the urine. Degradation was detected in kidney homogenates and a renal dipeptidase (dehydropeptidase-1) was found to be responsible for the metabolism. Here was irony indeed, for the compounds were stable to the microbial β-lactamases but susceptible to a mammalian β-lactamase, which rendered ineffective their use in urinary tract infections.

The Merck group then carried out a directed search for enzyme inhibitors and obtained a series of (Z)-2-(acylamino)-3-substituted-propenoic acids that were specific and competitive. The cysteinyl compound cilastin was selected as the preferred inhibitor for combination with imipenem in the commercial product Primaxin.

Production Process for Thienamycin. Production of thienamycin was a major problem also, since fermentation titres could not be increased sufficiently for commercial purposes as the initial synthetic routes to thienamycin yielded racemates and a biologically inactive double bond isomer predominated. In a brilliant piece of chemistry, Merck were able to devise a stereo-controlled total synthesis of (+) thienamycin. The problem of constructing three contiguous chiral centres was solved by using aspartic acid as starting material to generate the chirality of the specific enantiomer, and by the imaginative use of a carbene insertion reaction to produce the bicyclic nucleus.

6-Mercaptopurine, Allopurinol, Azathioprine and Methotrexate. The cell nucleus was the primary target for antineoplastic agents even before Watson and Crick published the structure of DNA. Gertrude Elion and George Hitchings, who pioneered many developments, began their work together in 1944. They were the first to show that inhibition of DNA synthesis in cancer cells could be achieved selectively by inhibitors of purine biosynthesis. Their synthesis and discovery of 6-mercaptopurine was a landmark event in medicinal chemistry. The compound was the first clinically useful drug for the treatment of acute childhood leukaemia, and a source from which flowed numerous other new medicines.

6-Mercaptopurine is rapidly oxidised by xanthine oxidase to form the inactive 6-thiouric acid, and in attempting to develop compounds that might block this inactivation, Elion and Hitchings discovered allopurinol (39), a potent inhibitor of xanthine oxidase. However, because it blocked the conversion of hypoxanthine to xanthine and uric acid, allopurinol became more widely used in the treatment of primary and secondary gout. Allopurinol acts as a 'suicide substrate' for the enzyme; it is oxidised at C-2 to give alloxanthine, which binds tightly and inactivates the enzyme. 6-Mercaptopurine was the touchstone to a further serendipitous discovery by Elion and Hitchings. In trying to decrease the rate of inactivation of the compound, they produced azathioprine. This was intended to act as a sustained release pro-drug form of 6-mercaptopurine. Although it disappointed in tumour therapy it was found to be useful as an immunosuppressant following transplant surgery.

The work of Elion and Hitchings concerned with inhibitors of dihydrofolate reductase (DHFR), provides a very good example of the selectivity that can be achieved when isoenzymes exist. These perform the same function in different tissues or different species, but allow different substrate specificities or inhibitor binding characteristics to be exploited with advantage. Thus in the case of DHFR, although the same enzyme reaction occurs in man and parasite, the forms of DHFR are sufficiently different that small molecule inhibitors like trimethoprim have great selectivity for the bacterial enzyme, whereas pyrimethamine has selectivity for the Plasmodium enzyme. Methotrexate, which inhibits human, bacterial and plasmodium enzymes to about the same extent, is of course itself a useful immunosuppressant. It was the first antimetabolite compound obtained able to produce remissions in leukaemia.

(39) Allopurinol

(40) Acyclovir

Acyclovir. The development of the anti-herpetic agent acyclovir transformed the chemotherapeutic approach to viral disease. Schaeffer and colleagues (1978) synthesised the compound (40), in which the acyclic chain mimics a part of the sugar moiety of deoxyguanosine, as a potential antimetabolite, based on earlier studies which had defined the binding domains of the enzyme adenosine deaminase. The Burroughs Wellcome group unexpectedly observed that the compound had potent activity against herpesviruses, particularly HSV-1 and HSV-2, and it soon proved very effective clinically against genital herpes and herpes infections of immunosuppressed patients. Elion (again) contributed to the understanding of the selective action of the compound, which becomes concentrated in the virus-infected cell. In the infected cell acyclovir is phosphorylated to the monophosphate by a virally-induced thymidine kinase. Selectivity is achieved since the host lacks this enzyme, and hence the monophosphate is not formed in uninfected cells. In the infected cell acyclovir is further phosphorylated to the triphosphate, which serves as a substrate for the virally encoded DNA polymerase. The substrate analogue gets incorporated into the growing DNA chain, but chain termination occurs since acyclovir lacks the 3'-OH required for the phosphodiester linkage.

Captopril. The clinical importance of blockade of the renin-angiotensin system emerged following the discovery and development of captopril (41), an angiotensin converting enzyme (ACE) inhibitor, first marketed in 1980. Peptides contained in the venom of the Brazilian viper Bothrops jararaca had been shown to potentiate the hypotensive activity of bradykinin and the pressor effects of angiotensin II. Ondetti and Cushman at Squibb synthesised some of these peptides that blocked the conversion of AI to AII, including the nonapeptide SQ-20881, teprotide. This ACE inhibitor was the first to be tested clinically, and along with other peptide inhibitors, provided the structure-activity base for their subsequent development of captopril, the first clinically useful, orally-active inhibitor.

Their design of this small non-peptide inhibitor was based on a hypothetical model of the enzyme active site. ACE is a zinc-containing peptidyldipeptide hydrolase which preferentially cleaves the terminal dipeptide from AI (and bradykinin), and they modelled the enzymatic reaction using the digestive enzyme carboxypeptidase A. This enzyme had been shown to be potently inhibited by the L-isomer of benzylsuccinic acid. This knowledge enabled them to predict that succinyl-amino acids would be accommodated in the active site of the ACE enzyme. Substitution of sulphydryl for the carboxyl function increased potency, and the (R) stereochemistry of the methyl satisfied the requirements for an L-amino acid in this region, eventually leading to captopril. The Merck group subsequently developed inhibitors, e.g. enalapril (42), the current market leader, which binds to an additional hydrophobic site on the enzyme.

Natural Products

In recent years there has been renewed interest in natural product sources and in the enormous diversity of secondary metabolites that can be obtained from bacteria, fungi and plants. Natural products have yielded both new drug entities (e.g. cyclosporin, mevinolin) and leads for derivatisation and development (e.g. asperlicin, avermectin, FK-506, staurosporine, calphostin, monobactams).

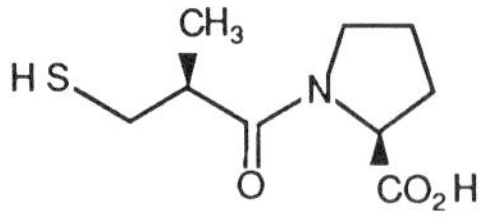

(41) Captopril

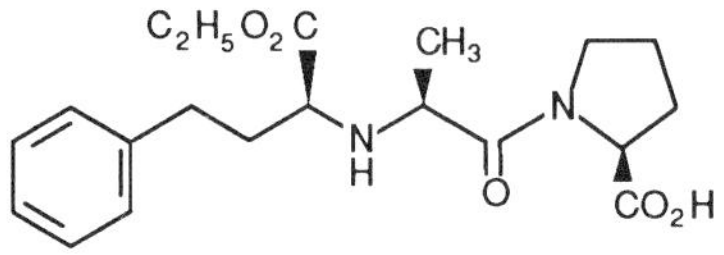

(42) Enalapril

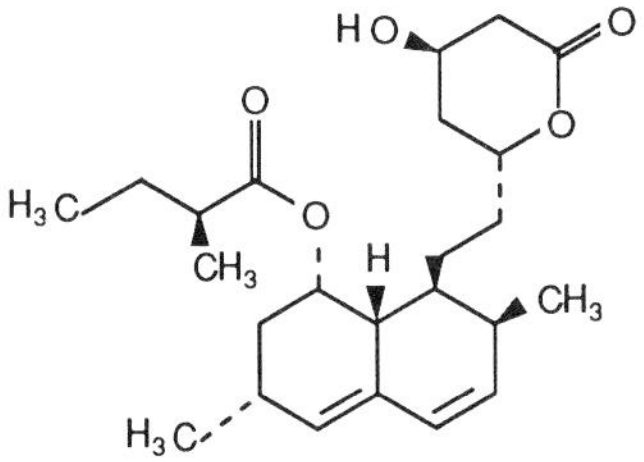

(43) Mevinolin

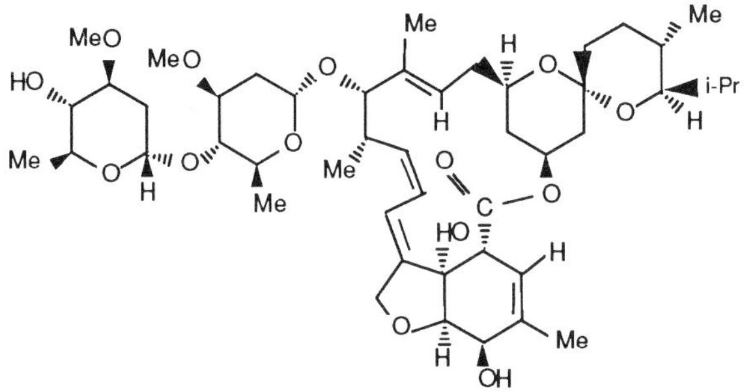

(44) Ivermectin BIB

Mevinolin. The discovery of mevinolin (43) may come to be recognised as a milestone in the drug treatment of one of the major risk factors of coronary artery disease, hypercholesterolaemia. HMG CoA reductase is the rate limiting enzyme in the biosynthesis of cholesterol. Potent competitive inhibitors of the enzyme were discovered in natural product isolation programmes at MSD. Compactin was obtained from cultures of Penicillium citrinum and P. brevicopactum, while the more potent mevinolin was obtained from Aspergillus terreus.

Ivermectin (44) is the 22,23-dihydro derivative of the natural product precursor avermectin B1, a fermentation product of Streptomyces avermitilis, which was isolated by the Merck group from a soil sample from Japan. The antihelminth activity of avermectins was discovered using a nematode screen, and because of their activity against both ecto- and endoparasites they have been called 'endectosides'. Ivermectin itself has a wide variety of veterinary uses and it may also be valuable in man for the treatment of river blindness. The mechanism of action of these macrolides appears to depend on interrupting neurotransmission through stimulation of GABA-mediated chloride ion conductance. This leads to selective toxicity for the arthropod and nematode parasites, which utilise GABA in peripheral neurotransmission; in mammals GABA-mediated mechanisms are restricted to the CNS which ivermectin does not readily access.

The Natural Product Discovery Process. This now usually involves the acquisition of organisms from very diverse habitats, grown under various conditions, from which large numbers of samples can be applied to high throughput mechanism-based screens. Productivity of the process becomes largely a function of the novelty and numbers of screening tests and organisms employed. Streptomyces sp. and fungi have been the most profitable sources thus far. An increasing number of new high capacity screens (>1000 assays/week) for enzyme inhibitors, receptor ligands and immunoregulatory agents have been added to those already developed for screening anti-infective agents. Examples of natural products obtained in this way include: cyclosporin A (immunosuppressant) from Trichoderma polysporin, FK-506 (immunosuppressant) from Streptomyces tsukubaensis; calphostin (PKC inhibitor) from Cladosporium cladosporoides.

Isolation, purification and structural characterisation of the biologically active compounds present in fermentation broths has been facilitated by advances in HPLC, mass spectrometry and high resolution NMR. In the Natural Products Discovery Department at Glaxo, known compounds are now rapidly eliminated and novel compounds characterised in a 2-stage procedure. In the first stage, crude biologically-active samples are chromatographed using automated HPLC and the fractions bioassayed. Without further purification, the active fractions are subjected to LC-MS with on-line diode array UV analysis. Compounds which match with known natural products (on the MS/UV natural products database) are eliminated unless they demonstrate a novel mode of action. In the second stage, samples containing potentially novel substances (often obtained after large scale fermentation, 50-500l) are subjected to preparative HPLC and/or countercurrent chromatography to provide sufficient material (ca. 5 mg) for full structural elucidation by mass spectrometry and NMR.

3 THE MEDICINAL CHEMISTRY INDUSTRY TODAY

The Costs of Drug Development. 150 years ago it was possible for Morton to demonstrate the anaesthetic properties of ether in patients within weeks of his discovery, and for supplies of the product to be made available without regulation or constraint other than that of the production process itself. Today things are different. For each new drug to reach the market, 10,000 compounds may have been prepared and £100m spent over 10 or more years in getting it there. A product needs to achieve annual sales of £50-100m to be profitable. Professor Trevor Jones recently suggested that to get an adequate return on $100 m spent on R & D, one must develop a drug with pre-tax earnings of $450m, equivalent to $1100m total sales revenue.

Problems for the Industry. The Centre for Medicines Research has found that since the 1960s the time between a drug patent application and marketing has increased 4-fold to an average of 12 years in the 1980s. The pre-clinical time (2 years) has remained about the same, but the clinical development and regulatory periods have greatly increased. This has led to a great erosion of effective patent life of a product. Other factors are operating presently to reduce margins, including: increased costs of R & D and marketing, pricing regulation, competition from generic drug prescribing and illicit importing. Companies are seeking survival strategies and these include mergers, internationalisation of R & D, strategic alliances with academe and biotechnology companies, but some extension of the effective patent life of products would be an important means of ensuring proper return for the enormous investment required in new drug development.

Impact of Medicinal Chemistry Products. The Medicinal Chemistry Industry has contributed significantly to the changes brought about by improved medical knowledge and health care. In the developed nations, infectious diseases of the young have lost their terror and infant mortality has decreased very substantially. In the III World however, infectious diseases remain extremely common and a challenge to healthcare provision. Effective methods of prophylaxis or active treatment have facilitated the expansion of travel. Other drugs and chemicals have had great impact on social, environmental and economic affairs. Thus the availability of chemical means of contraception, in particular the various oral contraceptive steroids, has had enormous impact on birth control methods and changed the social condition, status, health and economic outlook of women in societies.

Future prospects. Medicines have improved the quality of life for patients (e.g. pain-free surgery, control of diabetes) or reduced morbidity and mortality (e.g. from stroke and renal disease due to hypertension). And yet how much still has to be achieved. Little impact has been made on many diseases of the central nervous system (e.g. Alzheimer's, Parkinson's, schizophrenia), chronic bone and joint disorders (rheumatoid arthritis, osteoporosis), the commonest cancers (lung, breast, stomach, colon, prostate). Even in the cardiovascular and infectious disease areas, where greatest progress has been made, ischaemic heart disease still devastates and AIDS is spreading quickly.

Over the last decade, great progress has been made in establishing the regulatory and integrative roles of endogenous peptides and proteins in a whole variety of biological processes. Neuropeptides, interleukins and many other peptide species have been discovered, synthetic and analytical techniques have advanced enormously, and molecular biology and genetic engineering have enabled complex proteins to be produced quite readily. A whole new range of disease targets have been recognised for therapeutic intervention: on cell surfaces, at receptors, on enzymes, on nucleic acids. However, proteins and peptides are not good drugs, since they are unstable in biological fluids, are rapidly eliminated and transported poorly across membranes. As medicinal chemists we have been presented with the most exciting and difficult challenge - how to design bioavailable drugs that can fulfil the potential revealed by the new biology.

SELECTED REFERENCES

Baumberg S., Hunter I.S. and Rhodes P.M. (Eds.). 'Microbial Products: New Approaches'. Cambridge University Press. 1989.

Demain A.L., Somkuti G.A., Hunter-Cevera J.C. and Rossmoore H.W. (Eds.). 'Novel Microbial Products for Medicine and Agriculture', Elsevier, New York. 1989.

Fieser L.F. and Fieser M, 'Steroids', Reinhold Publishing Corporation, New York. 1959.

Hitchings G.H. and Elion G.B., Cancer Res., 1985,45,2415.

Lis Y. and Walker S.R., Pharm. J., 1988,240,176.

Porter R., 'Disease, Medicine and Society in England 1550-1860'. Macmillan, London. 1987.

Roberts S.M. and Price B.J., (Eds.). 'Medicinal Chemistry: The Role of Organic Chemistry in Drug Research', Academic Press, London. 1985.

Sneader W. 'Drug Discovery: The Evolution of Modern Medicines', John Wiley & Sons, Chichester. 1985.

Walker B.C. and Walker S.R. (Eds.). 'Trends and Changes in Drug Research and Development'. Kluwer Academic Publishers, London. 1988.

Walpole C.S.J. and Wigglesworth R. 'Enzyme Inhibitors in Medicine'. Natural Product Reports 1989,6,311.

Food

Fertiliser Progress 1841–1991: A Review of the Development of Mineral and Organic Fertilisers

Michael Dewhurst

HIGHFIELDS, HIGHFIELDS WAY, FRANCE LYNCH, STROUD, GLOUCS. GL6 8LZ, UK

FERTILISER PROGRESS 1841-1991

During the course of the past 150 years the concept of using supplementary mineral nutrients to feed plants to feed people has expanded. It has progressed from the ideas of a handful of scientists and early industrialists to a massive worldwide industry producing more than 150 million tons of fertiliser nutrients used to boost crop production.

Thomas Malthus stated that population growth would outstrip food supply. He died in 1834 just before the start of fertiliser manufacture and other technology that has helped to avert the total onset of his prediction. In developed countries agricultural science has given farmers the ability to supply more than enough food needed by people in industrialised society. In poorer countries millions of people, unable to fully utilise the benefits of science, technology and commerce, are suffering the calamity that Malthus foretold.

This progress report on fertilisers takes account of scientific agricultural and industrial advances made since the mid-19th century.

Fertilisers and Population Growth

In the 1840s the total world population was probably about 1.0 billion people, today FAO estimate that over 5 billion now occupy the planet. The problem of feeding people in many parts of the world is just as urgent today as it was in the 19th century.

YEAR		ESTIMATED WORLD POPULATION
1650	...	500,000,000 estimate
1850	...	1238,000,000 200 years to double
1950	...	2517,000,000 100 years to double
1987	...	4996,000,000 about 40 years to double
2025	...	9000,000,000 about 35 years to double (proj.)

Ref:UN(1989) Prospects of World Urbanization Pop. Studies no 112

During the 1800s rapidly expanding and increasingly urbanised populations in industrial Europe and North America demanded more food. Earlier generations of farmers had more or less kept pace with demand by using new land, animal manures and crop residues to supply plant foods. Food production from mixed organic farming could no longer sustain people who were moving from the countryside to the towns.

It is widely accepted that the fertiliser industry started in the 1840s and from then to the early 1900s there was a slow but steady increase in demand for fertilisers. Superphosphates were produced, guano and sodium nitrate were imported from South America and industrial by-products such as ammonium sulphate and slag were found to be valuable fertilisers.

Towards the end of the century it appeared possible that the supply of nitrogenous manures would run out. Sir William Crookes spoke of the increasing concern that wheat production would suffer unless natural supplies of nitrogen were supplemented by manufactured fertilisers. Within a few years of his famous address to the British Association, 3 different methods of synthesising fertilisers from atmospheric nitrogen were developed.

An estimate of world supplies of fertiliser by Sir John Russell in 1933 indicates slow growth of fertiliser use through the period of agricultural depression. (7) Fertiliser technology advanced considerably during the period from the 1930s to 1950s with the invention of better chemical sources and processes and improved particle formation and product packaging. The worldwide growth of the industry since 1950 has been very rapid. It is based on the availability of natural gas as a feedstock for the production of nitrogen to help fulfil the improved yield potential of staple crops.

Manufactured Fertiliser Consumption (World)
million tons N + P_2O_5 + K_2O

1650	small amounts of organic waste and ground bones
1850	est. 0.03million tons (JMD)
1950	... 20 million tons (FAO)
1987	... 132 million tons (FAO)
1991	est.144.8 million tons (FAO/UNIDO/World Bank/Ind.Gp)

The distribution of plant nutrients for agriculture, like human food, remains inequitable. The average amount of fertiliser used is about 30kg of nitrogen, phosphate and potash per person. Industrialised societies utilise 60kg per head whereas in Africa no more than 6kg per person are available.

Fertilisers, Soils & Plants

By the early 1840s a number of scientists had developed and tested theories of plant nutrition. Justus Liebig brought these theories into sharp and critical focus with his address to the British Association,

'Organic Chemistry and its Application of Agriculture and Physiology'. (1)

When it was published it stimulated much discussion and Liebig is often accredited with being the first to suggest treating bones with sulphuric acid. By changing the form of calcium phosphate to a more soluble form he found that crops

'grew with much vigour'.

Liebig acknowledged the work of other scientists. The idea of improving the effectiveness of bones and mineral rock phosphates appears to have occurred to several innovators at about the same time, since commercial production of superphosphates started in several places. The following people are known to have opened 'manufacturies' in the 1830s & '40s

Kohler in Bohemia, Lawes in England
Escher in Moravia, Murray in Ireland

Many of the 'agricultural improvers' of the day tried new ideas. Jean Baptiste Boussinghault, a French engineer and explorer, appears to have been among the first of a number of innovators to follow the work of de Saussure by applying fertiliser nutrients to crops in the same ratio as occurred in plant ash. Boussinghault set up scientific trials to test different mineral salts as fertilisers. His experimental farm in Alsace was followed by others, notably at Rothamsted where Lawes and Gilbert created the foundations of modern soil science and plant nutrition.

This movement, emulated in Europe and North America, led to a worldwide network of experimental stations. Today the scientific team at Rothamsted continue to illuminate the complex relationship between soils and plants. Unfortunately work at the farm at Bechelbronn ceased during the Franco-Prussian war of 1870.

Fertilisers and Crop Yields

The world's food supply depends on crop yield and cultivated area. Crop yields are limited by a number of factors including nutrient and water supply. They are also constrained by the capacity of particular plant varieties to produce high yield. Plant breeding has contributed greatly to increased crop yields and therefore to the plant's requirement for nutrients. In 1843 Liebig described the best wheat crops yielding 5.6 t/ha of straw and 2.8 t/ha of grain and for nearly a century average European wheat yields were below this level.

In the 1960s and 70s there was a great leap forward in plant breeding. Nowadays some wheat crops in Europe will top 10 t/ha of grain with 7 t/ha of straw. Over the 150 year span grain has increased by three or four fold. On a world scale increase in grain production from each unit of land is equally dramatic, though starting from a lower base.

TRENDS IN CROP YIELD IN EUROPE (t/ha of harvested crop)

	Wheat (UK)	Sugar Beet (France)	Maize (France)
1900	2	26	1.1
1950	2.6	34	2.2
1970	4.2	45	5.1
1988	6.2	60	7.2

Agriculture & Fertilizers (1990) A Norsk Hydro report

Progress in Fertiliser Production

The major nutrients in industrially produced fertilisers are mineral forms of nitrogen, phosphorus, and potassium. In addition sulphur, as well as being a plant nutrient, has a pivotal role in phosphate fertiliser manufacture. Sulphuric acid treatment changes phosphate in natural rocks to chemical forms that can be more easily taken in by plants. Magnesium and sodium, associated with potash production, are also important fertiliser nutrients. The major nutrient calcium is normally traded locally as limestone or chalk and there are specialist products such as micro-element fertilisers, slow release fertilisers and organic fertilisers. Most fertilisers are usually applied to plants as solids but are sometimes applied as liquids.

Phosphates Liebig was not the only man of 'assiduous inquiry' working on the development of fertilisers. In England, the young John Bennet Lawes designed a process for improving the effectiveness of crushed bones as a fertiliser. In Ireland, Sir John Murray followed a similar course. They both, coincidentally, registered a patent on the same day, May 23rd, 1842. Later Lawes changed his patent to the production of 'superphosphate of lime' from phosphorites. He used locally occurring nodules of calcium phosphate known as 'coprolites', as well as bones, in his 'manufactory' at Deptford. Lawes finally bought the Murray patent and it is usually acknowledged that the Lawes Chemical Company developed the first purpose-built fertiliser factory in Britain.

Within the next ten years about 18 fertiliser factories were established in the British Isles for the production of superphosphates. By 1870 there were about 80 factories producing this type of fertiliser. Most of them were situated at ports and established an excellent export trade, as well as shipping to smaller ports around the country. Developments in phosphate fertilisers continued with importation of better grades of apatite and the development of continuous manufacturing processes. Some of the companies established then, are still in business today.

Basic Slag In 1878 Thomas and Gilchrist introduced a method of removing the phosphorus impurity from pig-iron. The by-product 'basic slag' proved to be a valuable fertiliser. German farmers were the first to recognise the agronomic value of basic slag and for many years this form of fertiliser was widely used there. In Britain, Wrightson and Munroe showed how the application of basic slag improved pasture land and it was popular with British farmers for nearly a hundred years. Changes in steel making methods in the late twentieth century have virtually removed all basic slag from the fertiliser market.

Chemical forms of phosphate fertiliser continued to be improved with the development of phosphoric acid which allowed the manufacture of concentrated superphosphate and the advent of the ammonium phosphate group of chemicals. Most phosphates are produced close to the mines in North Africa, USA, Finland and Russia where apatites are extracted from the earth's crust.

Today triple or concentrated superphosphate(46 P_2O_5) and ammonium phosphate mixtures, known commercially as DAP ('diammonium phosphate'46 P_2O_5) both in high grade granular form, dominate world trade in phosphate fertilisers. Some granular ground rock phosphates are also used.

Nitrogen The first great surge in interest in what was then termed 'artificial fertiliser' was created by the importation of guano around 1840. Today we would call this dried dung a semi-organic fertiliser and it came as a revelation to many farmers. When someone in a locality tried it, the results were talked about in markets and wherever farmers met. In the space of a few years imports grew from a few hundred tons to many thousands of tons. By the 1870s the valuable deposits of seabird dung, containing upto 16% nitrogen, 14% phosphate and 3% potash, were nearly exhausted. Demand for other nitrogen fertilisers increased as the 'miraculous' guano became adulterated and spoilt, and lost its power to boost yields.

"'I got a ton and a half of guano at Bradley's in High Street,' said the Archdeacon, 'and it was a complete take-in. I don't believe there was five hundred-weight of guano in it'." Barchester Towers (1857) Anthony Trollope.

Sodium nitrate from the caliche deposits of South America and ammonium sulphate, a by-product of coal gas production, rapidly became the most important mineral nitrogen fertilisers available to farmers.

At the close of the nineteenth century the world was dependent on nitrates from Chile and ammonium salts from coal. Other important, mostly non-commercial, sources of nitrogen were human and animal waste, legumes and by-products from food processing. There were serious doubts about whether these sources would be sufficient to meet agricultural needs for nitrogen fertilisers. In 1898 Sir William Crookes, portrayed a vivid picture of famine among the

"wheaten bread eating races of the earth."
Within a short time of his appeal for

"starvation to be averted through the laboratory", development of industrial atmospheric nitrogen fixation was under way. The electric arc process for nitrate production was accomplished in Norway, based on the Berkeland and Eyde furnace calcium nitrate produced in this way was offered for sale in the early days of the new century. In addition, German scientists, notably Caro and Frank made calcium cyanamide and by 1905 this new fertiliser was also on the market.

Important though these advances were in the progress of fertiliser manufacture, the major step forward was made by Fritz Haber who published his ideas on ammonia synthesis in 1905.

Full scale industrial ammonia production was achieved in 1913 under the guidance of Carl Bosch, the chief chemical engineer for Badische Anilin und Soda-Fabriken, BASF. This development was spurred on by massive grant aid from the German government who were aware that the supply of nitrates for explosives could become critically short in the event of war.

In the intervening 80 years or so since the advent of ammonia production, materials produced by chemical engineers have grown to be virtually indispensable to modern man.

In 1880 Ludwig Mond observed that:
"Amonia synthesis is indeed the beginning of a new era of chemical technology, leading to a widespread application of high pressures and temperatures in many chemical processes."

Ammonia synthesis rapidly overtook other methods of nitrogen fixation. Most modern fertilisers, such as ammonium nitrate, urea and ammonium phosphates, are manufactured from ammonia feedstock. Today economic production of ammonia is directly related to abundant supplies of natural gas. The development of the world's gas fields coincided with the great upsurge of demand for nitrogen fertiliser. The plant breeding programme known as The Green Revolution produced new crops with increased yield potential that required more fertiliser.

In the last 30 years major improvements have been made to the chemical stability and physical form of nitrogen fertilisers. Granular, prilled and fluid forms of ammonium nitrate and urea help farmers to fertilise their crops with precision.

Potash The development of potassium fertilisers occurred later than the other two major nutrients. Farmyard manure and plant residues were the main source of potash during the nineteenth century, and on some soils these sources supplied adequate potassium to crops. As areas of wheat growing expanded on the lighter bodied soils, that naturally contain less potassium, Liebig's famous Law of the Minimum began to operate for potassium. It became clear that potash fertiliser was required to achieve optimum yields.

In the late 1860s in Britain, Augustus Voelcker introduced small quantities of potash from the Stassfurt mines in Germany. He showed that crops grown on light sandy soils were responsive to potash fertiliser.

Thomas Brown imported 200 tons of kainit into Kings Lynn in 1870 and Norfolk farmers quickly realised the advantages of this crude potash salt for use on mangolds. In the twentieth century this combined sodium and potassium fertiliser has greatly benefited sugar beet.

While a traditional trade was becoming established in East Anglia the rest of Britain remained sceptical about the worth of potash. One merchant F.W. Berk gave away much of his imported potash 'to encourage the making of trials'.

The Potash Syndicate, set up in 1890, prompted the use of potassium salts. It was probably the first British, nationwide commercial service to advise on the proper use of fertilisers.

Augustus Voelcker established an extremely successful private consultancy on the analysis fertilisers and gave much advice on the proper use of fertilisers, the continuity of advice was maintained by his son John Augustus Voelcker. The family firm Dr Augustus Voelcker & Sons Ltd remain Honorary Consulting Chemist to the Royal Agricultural Society of England after 115 years.

In Europe, particularly in Germany, potash was used very successfully to bring light sandy soil into agricultural production. By the turn of the century over 1 million tonnes per annum of potash were taken from the Stassfurt mine.

Today modern potash fertilisers have hardly changed in their chemical make up. Kainit and sylvinite remain popular crude salts, while potassium chloride is the main source of nutrient for fertilisers. Still known widely as Muriate of Potash it is used as a powder or compacted and graded into precisely measured particle sizes for sale as a straight fertiliser or to add to mixtures known as compounds.

There are many deposits of potash rock in the world, occurring as evaporites and originating from the remains of ancient seas. In 1976 an important deposit in Cleveland, North Yorkshire was developed and this plant supplies Britain and Europe with potash and sodium salts.

Compound Fertilisers Mixed fertilisers were available from the beginning of fertiliser development. The best samples of guano, for instance, contained all three major nutrients. Many farmers found it difficult to know which mineral fertilisers they could safely mix together in spite of numerous text books containing tables on how to do it. Fertiliser merchants and manufacturers started to produce ranges of fertilisers with varying plant nutrient ratios. These proved to be commercial winners, since most farmers were quite happy that their supplier ran the risk of the whole heap setting as hard as rock. This practice involved sellers of fertilisers in the decisions about which ratios were best suited to crops and soils in the locality. Most companies were pleased to supply an information service for their customers, usually based on the opinions of experienced staff who could advise on the best mixtures.

A great advance came in the 1920s and 30s when BASF in Germany, and ICI in England, developed concentrated granular compound fertilisers. They soon became firm favourites with farmers, having two major advantages over powdered mixtures. Firstly the granular form greatly improved storage, handling and, above all, made accurate application by machine possible for the first time. Secondly granular compounds were made up of chemical salts that increased concentration of nutrients. One ton of CCF ('complete concentrated fertilizer'), launched commercially in 1932 and based on ammonium phosphate with the analysis 12.5%: N:12.5% P_2O_5 15% K_2O required 1.7 tons of powdered mixture to carry the same nutrient content. German compounds were equally effective but were based on nitrophosphates.

The next important progression came from America in the 1950s in the form of blended granular compound fertilisers. Developments in the concentration and improved physical forms of single nutrient and intermediate fertilisers allowed the blending or dry mixing of these materials. In a sense it was a return to the 'powder mix' but now the enhanced physical forms allowed virtually all the disadvantages to be overcome. In addition the new method allowed any ratio of plant nutrients to be tailor-made for individual customers at very competitive prices. This type of mixed-nutrient fertiliser introduced into Britain in 1968 is now very popular with farmers.

Fluid Fertilisers A substantial amount of fertiliser is applied to agricultural and horticultural crops in fluid form. Farmers have always valued effluent liquor from the 'muck heap' as a fertiliser. Manufactured fluid fertilisers were developed over the last 40 years. The major progress coming from the USA with local innovations made in other countries. Solution fertilisers contain all the constituent nutrients in dissolved form. Suspension fertilisers use finely ground solids suspended in water by means of dispersed clay particles. Both these types are applied through farm sprayers. In some parts of the world solutions are applied through irrigation systems and this process is known as 'fertigation'. Ammonia fertilisers are sold as anhydrous ammonia, mainly in America and also as aqueous ammonia. Both types are usually applied by injecting the chemical below the soil surface to avoid gaseous loss.

Progress in Farm Use of Fertilisers

There are many benefits of using fertilisers to grow high yields of good quality crops, to both individual farmers and to society as a whole. It is also perceived by many people that there are disadvantages to over-using fertiliser nutrients and allowing leakage into the environment.

It is every farmer's responsibility to look closely at the management of fertiliser inputs in order to make sure he is using fertilisers properly. The governments of developed countries, in particular, are now formulating codes of practice and enacting laws that make sure he does.

The modern grower is able to use fertiliser at the correct rate and at the right time in the growing season. When these decisions are made, fertilisers can be applied accurately to crops in gardens, glasshouses and fields.

Over the 150 years of fertiliser progress, chemical engineers have rendered an admirable service to agriculture, horticulture, forestry as well as gardeners and pot plant enthusiasts by producing a wide range of sophisticated products relatively cheaply in order to meet the needs of modern society. This includes the small minority that wish to return to the organic methods of the nineteenth century who are catered for by the modern processing of waste materials.

Mechanical engineers have also made a vital contribution by close cooperation with the fertiliser industry. Improved fertilisers have led to the development of machines that will accurately deliver the required level of nutrients to the exact place in or on the soil where plants can best utilise them.

Improved packaging has also played an important part in more efficient fertilising by keeping the fertiliser in a spreadable condition. The prevention of moisture ingress into what are, by definition, soluble salts has taxed the ingenuity of man throughout the 150 year progress. Watertight casks were the mode in the early years. Later came closely woven hessian bags that were cheaper but not so effective. Bitumen lined paper proved a big improvement, finally plastic materials appeared to be ideal but not totally hard-wearing and weatherproof. Recent developments in plastic film technology has led to the situation where fertilisers can be left outside. In some areas fertilisers are handled in bulk from factory to farm, sometimes the material is delivered directly to the field and spread by contractor in liquid or solid form.

Finally the modern descendants of Lawes and Liebig continue to shed light on the intricacies of the complex nitrogen and other nutrient cycles. The results of their work at research stations around the globe are extended to farmers by teams of advisers in most countries of the world.

The aims and objectives of the modern fertiliser industry are quite clear:

> To help to prevent the realisation of Malthus' predication of imminent starvation that affect 4/5ths of mankind living in the developing countries of the world.

and

> To help to achieve the food production objectives of the remaining one fifth of the world's population living in the developed countries without excessive nutrient enrichment of soils and water.

Glossary of Terms

Compacted particles, materials extruded through dies broken up and graded into 2-5 mm diameter.

Fluid fertilisers, aqueous solution liquid fertilisers, suspension fertilisers, maybe part solution or part solids suspended in water, aqueous and anhydrous ammonia.

Granular particles, roughly textured spherical particles, 2-5 mm diameter made by rolling a drying slurry in a rotary 'drum'.

Prilled particles smooth spherical particles, formed by 'spraying' molten material down a tower.

Further Reading

Justus Liebig, 'Chemistry in its Application to Agriculture and Physiology' 3rd Edn., Taylor & Walton, London, 1843.

J.Royal Agric.Soc.England 1847,**8**,226:1851,**12**,1: 1855,**16**,411

Comp. Rend. 1877,84,301: 85,1018: 1878,86,892

C.M.Aikman, 'Manures and the Principles of Manuring' Wm Blackwood & Sons, Edinburgh & London, 1894.

Standard Cyclopedia of Mod. Agric. & Rural Economy Vol. 1-12, The Gresham Pub. Co., London, 1910
A.D.Hall, 'Fertilisers and Manures' John Murray, London, 1918

E.J.Russell, 'Artificial Fertilizers in Modern Agriculture' MAFF, HMSO, 1933.

A.S.Barker, 'The Use of Fertilizers', Oxford University Press, Humphrey Milford, London 1935

E.J.Russell 'History of Agricultural Science in Great Britain' Allen & Unwin, 1966.

George Dyke, 'John Bennet Lawes: the Record of his Genius' Research Studies Press, Harpenden, 1991.

The Fertiliser Society - Proceedings 1947-1991 Peterborough, PE3 6GF.

Fertiliser Progress

Before 1840- Saussure makes detailed analysis of plant ash. Berzelius describes action of sulphuric acid on bones. 1812 H & T Proctor start manufacturing fertilisers. Davy publishes lecture, Elements of Agricultural Chemistry. Gas Light & Coke Co produce ammonium sulphate liquor. Wohler synthesises urea. Nitrate of Soda introduced. Boussinghault at Bechelbronn. Liebig's lecture. Lawes starts experiments at Harpenden.

1840 Liebig publishes 1st edn. of British Association paper.
1841 Boussinghault publishes Bechelbronn results.
1842 Lawes and Murray register for 'supers' production.

Fertiliser Progress (Continued)

1843	Lawes joined by Gilbert, Rothamsted established.
1843	Coprolites used by Packard at Snape.
1845	Sulphate of ammonia reintroduced by Fownes.
1851	Stassfurt mines start potash mining on commercial scale.
1858 - 1870	Great Phosphate Rush - coprolites & Guano come & go.
1864	Voelcker introduces potash to Britain.
1898	Crookes appeal for N fixation process.
1899	Frank & Caro - produce calcium cyanamide.
1900	Bradley and Lovejoy produce calcium nitrate at Niagara but commercial success goes to Berkeland & Eyde in Norway.
1905	Haber publishes paper on ammonia synthesis.
1912	Bosch completes factory for ammonia production.
1913	BASF industrial synthesis of ammonia starts.
1914	Chilean Nitrate Co. says atmospheric N not needed.
1917	Monoammonium phosphate first produced in USA.
1922	First industrial synthesis of urea in Germany.
1923	Ammonia production started at Billingham.
1927	Calcium ammonium nitrate produced at Billingham.
1927	Granular nitrophosphate based compounds made in Germany.
1928	Ammonium nitrate first used as fertiliser in USA.
1930	Granular ammonium phosphate compounds made at Billingham.
1934	First granular superphosphate produced at Cliff Quay.
1935	Advances in phosphoric acid production in Canada.
1941	Further advances in economic production of phos. acid USA.
1950	Advances in stainless steel vessels aids acid production.
1951	Stengel process reduces risks from ammonium nitrate.
1951	Prilled ammonium nitrate first produced in USA.
1953	Granular triple superphosphate produced at Immingham.
1950s	Fluid fertilisers popular in USA (34% of N consumption)
1955	Bulk blending started in America.
1950s	Most fertiliser sold in 1 cwt paper sacks.
1962	Fertiliser solutions introduced to UK.
1962	Plastic bags introduced in Britain.
1965	Ammonium nitrate prills produced at Avonmouth.

Fertiliser Progress (Continued)

1967 Anhydrous ammonia fertiliser fails to develop in UK.
1968 Bulk-blending introduced to UK.
1969 Ammonia and fertiliser production started at Ince.
1975 Intermediate bulk containers (IBC) introduced to UK.
1977 Suspension fertilisers introduced to UK.
1980s Increase in popularity of bulk blended fertilisers in UK.
1983 IBC with integrated pallet introduced.
1989 All weatherproof IBC introduced in UK.

A Century of Crop Protection Chemicals

D.A. Evans and K.R. Lawson

ICI AGROCHEMICALS, BRACKNELL, BERKSHIRE RG12 6EY, UK

1 SUMMARY

Farmers and growers world-wide rely very substantially upon crop protection chemicals to help them meet the ever-increasing demand for food and materials such as natural fibres. The consumer continues to seek higher quality and more variety in produce. In post-war years, the industry has been able to play its part in meeting these demands by the continuous introduction of novel crop protection chemicals into the international market place. Today, there is an effective weedkiller, insecticide or fungicide to combat almost every significant problem faced by the modern farmer.

The paper begins with a factual review of the size and shape of the modern crop protection industry and the sectors within which it operates. The industry is overwhelmingly science and technology-based and the contribution of research and development to its success is discussed. The next topic is a description of how crop protection chemicals have evolved over the past hundred years or so from low-potency non-specific agents to highly active, selective and safe treatments. The methods of research and development leading to modern day agrochemicals will be described with particular reference to case histories of some key scientific discoveries. Examples illustrating the comparative success of random synthesis, analogue chemistry, exploitation of natural product leads and rational design will be provided in detail.

We conclude by stating that the industry has achieved a remarkably high degree of technical success.

In each of the sectors of fungicides, herbicides and insecticides, chemical treatments are available today which act effectively at application rates as low as tens of grammes per hectare. Notwithstanding the success of the products of the industry to date, we believe that there are many significant opportunities for new crop protection chemicals in the future. We can still improve substantially upon existing treatments, particularly in the areas of crop safety, inherent activity and flexibility in use. The onset of widespread resistance to many classes of insecticides and fungicides (and herbicides in recent years) provides a continuous requirement for chemicals with novel modes of action. In order to meet the demands and expectations of the public, farmers and the regulatory authorities, the industry is dedicated to the provision of new chemicals which are environmentally-benign and free of hazards to operators. The fundamental research base of the industry provides a sound footing for the work required to achieve these objectives, and we are confident of its success.

2 CROP PROTECTION

The world market for crop protection chemicals in 1989 was estimated at 24 billion dollars (US). Whereas this represents a 7.5% increase over 1988 in US dollar terms, the real growth was close to zero if currency movements and inflation are taken into account. The top fifteen agrochemical companies (1990 turnover) which operated in this market are listed in Table 1 together with sales data[1]. Three major sectors account for over 90% of the total market. The biggest sector is herbicides which makes up *ca* 44% of the total. Insecticides and acaricides come next with 29% of the market. Fungicides, *ie* chemicals used to control plant diseases caused by pathogenic fungi, account for 21% of the world market. The remaining 6% was comprised largely of chemicals which regulate plant growth and those which control nematodes in the soil. The areas of the major crop plantings worldwide are shown in Table 2 (for 1989-90)[2].

The global figures, however, hide enormous variations from region to region and country to country. For example, in France, fungicides are very much more important (30% of the market in 1990) than in North America, where herbicide usage (65%) outstrips massively the fungicide sector at only 7% of the market. In India, the insecticides sector is of greatest importance (70%). In Europe, treatments for small grain cereals such as

Table 1 Top 15 Agrochemical Companies (1990)

COMPANY	RANK	SALES ($M)
Ciba-Geigy	1	>2500
ICI	2	
Bayer	3	2000-2300
Rhone-Poulenc	4	
DuPont	5	
Monsanto	6	1500-2000
Dow-Elanco	7	
Hoechst	8	1300-1500
BASF	9	
Schering	10	
Sandoz	11	700-1300
American Cyanamid	12	
Shell	13	
FMC	14	400-700
Sumitomo Chemical	15	

wheat and barley take a large share of the market, whereas in Japan, rice dominates all other crop sectors. Similarly, the intensity of usage of agrochemicals varies enormously from country to country. Thus, in the USA, over 90% of the maize (corn) acreage is treated with herbicides whereas the comparative figure in Brazil is

Table 2 Major Crop Plantings Worldwide 1989-1990 (Million Hectares)

CROP	1989	1990	% CHANGE
Wheat	225.5	231.1	+2.5
Coarse Grains	322.9	322.0	-0.3
Rice	146.3	146.1	-0.1
Total Grains	694.7	699.2	+0.6
Cotton	32.0	33.5	+4.7
Soybeans	57.8	55.0	-4.8
TOTAL	**784.5**	**787.7**	**+0.4**

only *ca* 10%. Given the large differences in agricultural practice from country to country, it is not surprising that a large number of agrochemicals are required to meet farmers' needs, and presently over 500 individual chemicals are available either as single treatments or in many cases as mixtures.

The impact of crop protection chemicals on crop yield is indicated by Figure 1[3]. Whereas yield increased little over several decades prior to the 1940's, a dramatic increase has occurred since, coincident with the introduction of modern herbicide usage. However, it should be noted that the application of inorganic fertilizers, planting of improved cultivars and improved husbandry have all played their part in achieving better yields.

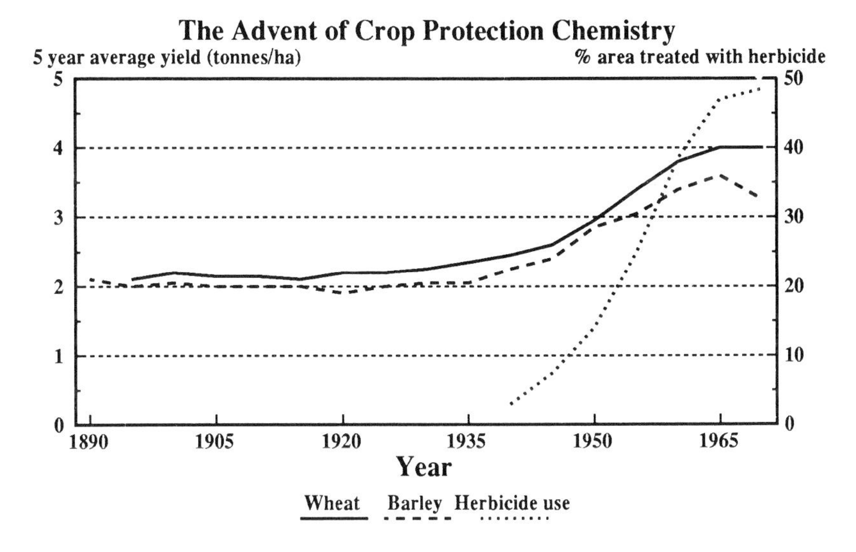

Figure 1

3 CROP PROTECTION PRODUCTS

In broad terms, there are seven major crop areas which contribute to the market for agrochemical treatments. These are maize, soybeans, small grain cereals (*eg* wheat and barley), rice, cotton, vines, and fruit and vegetables. Accordingly, most agrochemical companies share the same targets. This has led to a highly competitive situation in which the continued discovery and development of novel and effective crop protection compounds is critical for survival in the industry. Additionally, patentability is very important in order to secure a monopoly on a particular crop protection product and to exclude generic manufacturers who have not borne the high costs of research and development. These costs are estimated at around 100 million dollars (US) to take a new active ingredient from discovery to market place, over a typical timespan of ten years.

The characteristics of a successful new product will include:

- High levels of activity (down to grammes per hectare)
- High margins of safety to the treated crop
- Environmental safety
- Flexibility in use with respect to crop growth stage, temperature and weather conditions
- Compatibility with potential mixture partners
- Non-hazardous to operators and farmers

In the insecticide and fungicide sectors, there is presently wide-spread resistance to many classes of agro-chemicals. There is therefore a continuous need for the introduction of chemicals with novel modes of action in these areas. In recent years, resistance of weeds to herbicides has begun to appear, and in several cases the problem is assuming commercial importance.

4 HISTORY OF CROP PROTECTION CHEMICALS[4]

In 1985, the centenary of the introduction of the vine fungicide Bordeaux mixture was marked by an international conference held in France. The use of this mixture of copper sulphate and calcium hydroxide is generally taken to be the starting point of the modern agrochemical industry, although it should be stated that salt and industrial waste *inter alia* had already been in use as crop protectants for several centuries prior to this. Thus it can be said that the industry is just over a century old, explaining the title of this paper.

After 1885, progress in fungicide development was very slow and centred largely upon inorganic compounds. In the 1950's, the pace of development accelerated considerably and attention began to turn to organic compounds. The phthalimides (*eg* captan)(1)) provided broad spectrum protection of many crops against fungal attacks. In 1967, the introduction of benomyl (2) provided an exciting leap forward in fungal control. Benomyl exhibited the property of systemicity (*ie* movement within the vascular system of the plant) and was thus able to provide a curative action and to protect new growth.

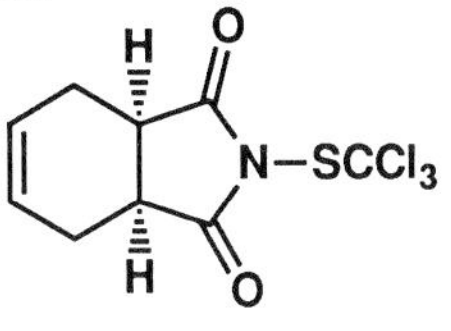

(1) 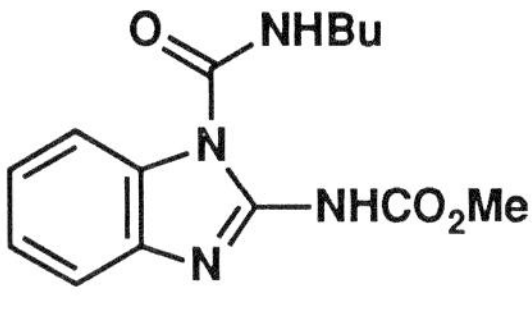 (2)

The major successes in recent years centre upon a very large series of commercialised chemicals which affect the ability of fungi to produce sterols, these being essential components of the fungal cell membrane. The largest family of these is known to inhibit the biosynthesis of ergosterol by inhibition of a C-14 demethylase. Many commercial companies have introduced highly successful products which show this mode of action (Figure 2).

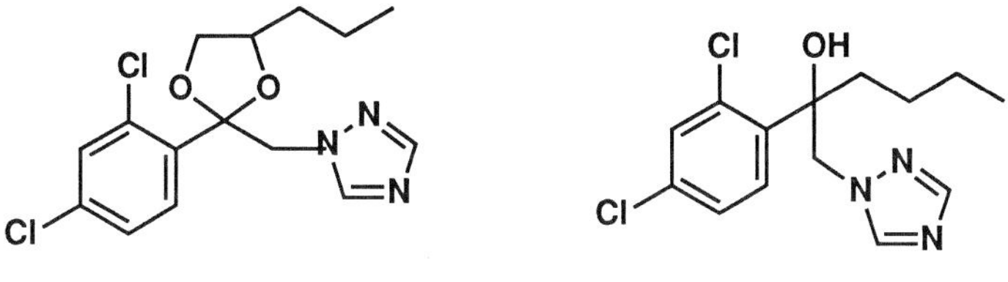

Propiconazole **Hexaconazole**

Figure 2 Examples of ergosterol biosynthesis inhibitors

Turning next to herbicides, the development of modern day treatments also stems in part from the use of Bordeaux mixture. It was observed in 1896 that the vine fungicide would also kill yellow charlock. The discovery of many other inorganic weedkillers then followed. The first organic chemical to be used commercially as a herbicide was 2-methyl-4,6-dinitrophenol (DNOC) (3), introduced in 1932. The key milestone in herbicide technology occurred in the early 1940's with the introduction of phenoxyacetic acid derivatives (mimics of

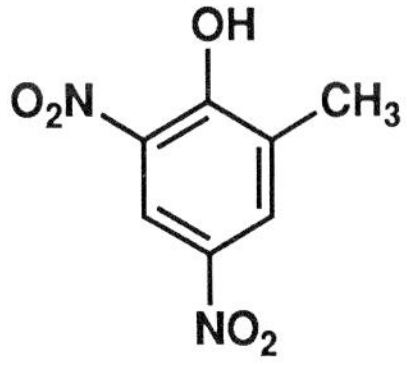

(3)

plant hormones) as selective weedkillers showing safety in cereal crops (*eg* MCPA (4)).

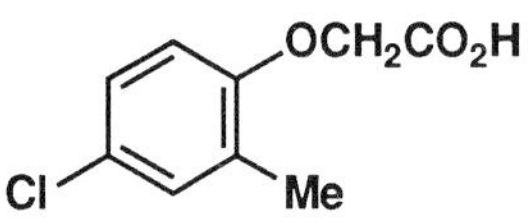

(4) 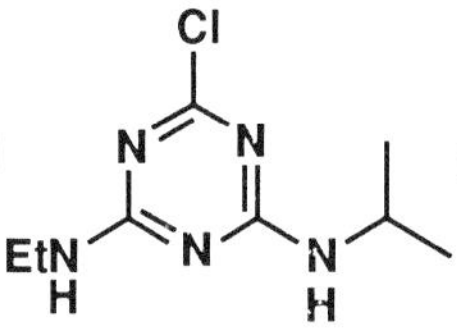 (5) 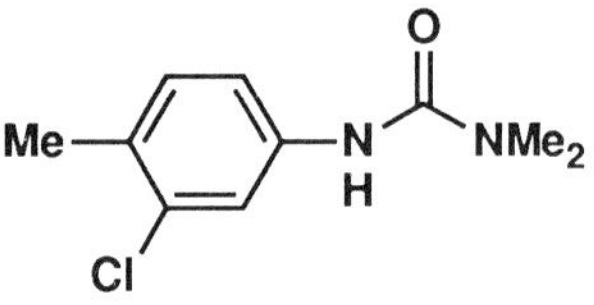 (6)

In the 1950's and 1960's, series of very important herbicides acting on the photosynthetic apparatus were progressively introduced *eg* atrazine (5) and chlorotoluron (6), and these were followed in the more recent past by a wide range of herbicides which provide the modern day farmer with treatments for almost all crop and weed situations.

In addition to highly selective chemicals, there were advances in total control of vegetation, paraquat (7) and glyphosate (8) (discussed in 6.1 below) being the most prominent examples.

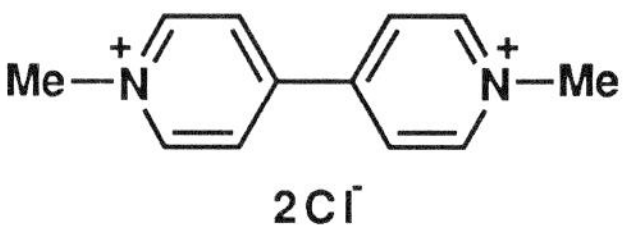

(7) 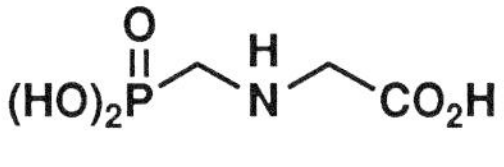 (8)

A major breakthrough occurred with the discovery of series of compounds which inhibited the enzyme acetolactate synthase, namely the sulphonylureas (*eg* chlorsulfuron (9)) and imidazolinones (*eg* imazethapyr (10)). Many of these compounds are very effective at 10 to 20 grammes per hectare treated, in marked contrast to the situation only a few year ago when it was common to apply agrochemicals at rates of kilograms per hectare.

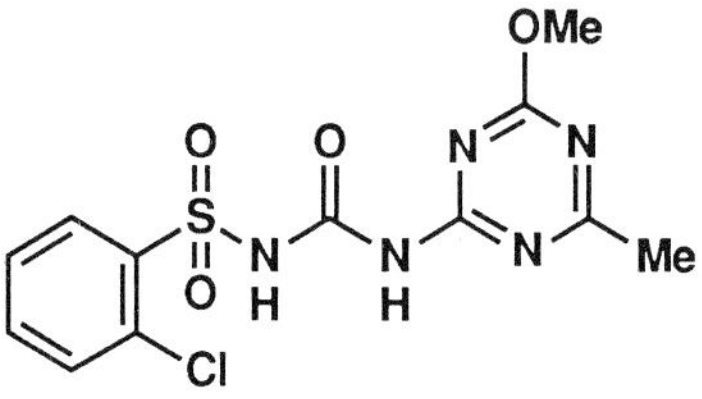

(9) 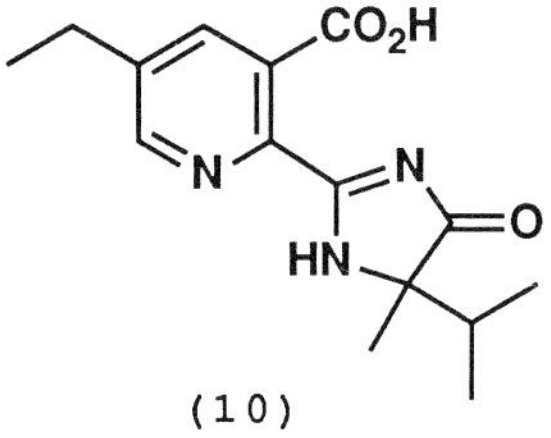 (10)

Finally, the insecticides present a very interesting picture of development. Prior to the last war, the unwelcome attentions of insects were combatted typically with inorganics (*eg* sodium fluoride) or with extracts of natural product obtained from plant sources (*eg Derris* extracts, and nicotine). During the last war, the introduction of DDT (11) revolutionised insect control.

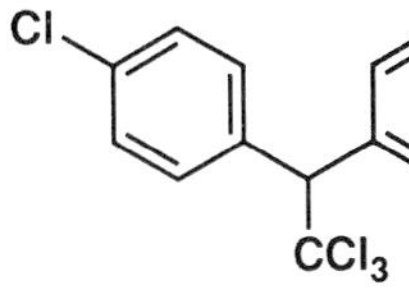

(11) 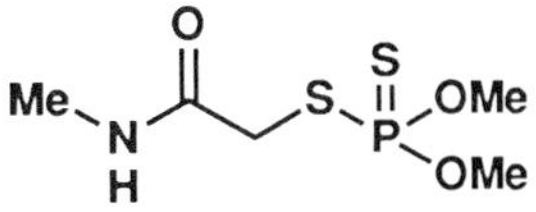 (12)

Organophosphate insecticides *eg* dimethoate (12) followed a few years later. These acted upon the insect nervous system as inhibitors of acetylcholinesterase and were quickly followed by the related carbamates, *eg* carbaryl (13), which were developed as agents less toxic to mammals.

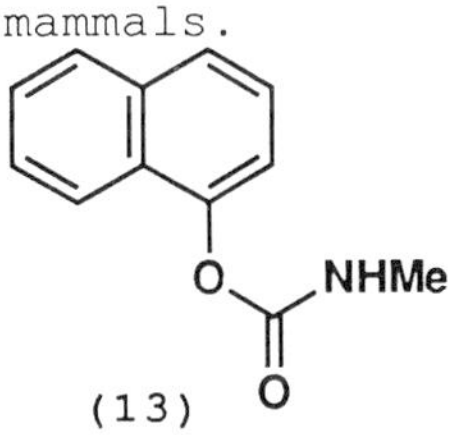

(13) 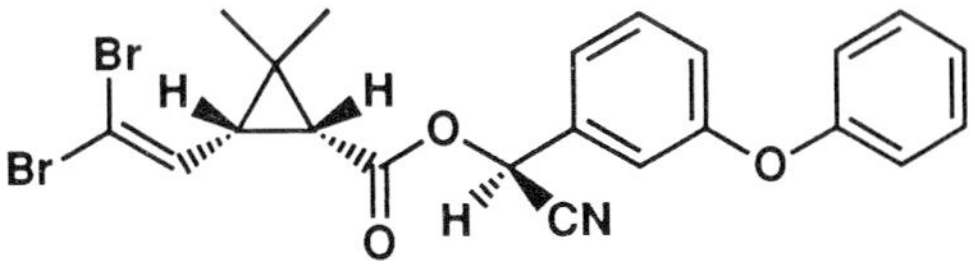 (14)

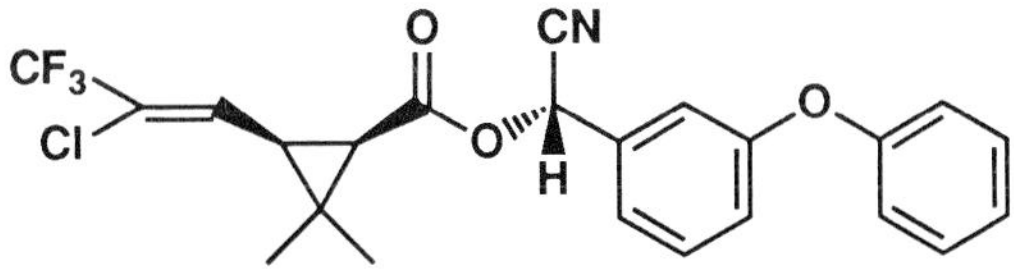

(15) racemic

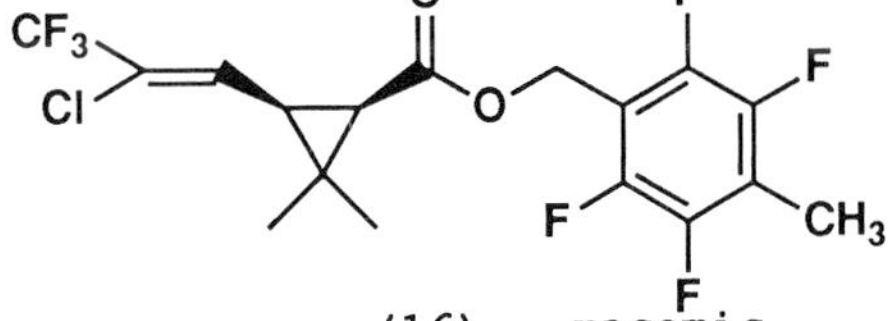

(16) racemic

The outstanding success of insecticide research was provided in the 1970's with the introduction of the synthetic pyrethroids which were derived in concept from the pyrethrins, natural product insecticides obtained from a chrysanthemum species. Deltamethrin (14) and λ-cyhalothrin (15) are typical examples of recent introductions which provide effective insect control at rates as low as 10 grammes per hectare. A significant advance has been the introduction of a pyrethroid (tefluthrin (16)) which is effective as a soil insecticide. This compound has a relatively high vapour pressure which, notwithstanding its high lipophilicity, allows movement and activity in the soil.

5 THE DISCOVERY AND DEVELOPMENT OF CROP PROTECTION CHEMICALS

The basic methodology for the discovery of new active compounds has remained in essence unchanged for decades. Novel chemical entities are obtained either by synthesis programmes or from natural sources *eg* by extraction of plants and micro-organisms. Biological activity is then detected by screening *in vivo* against a range of fungi, insects and plants.

It is not surprising that such screens form the cornerstone of research in every major agrochemical company, since they are cost effective, accurate, reproducible and can accommodate a relatively high throughput. In more recent years, the industry has moved towards a more targeted research approach in which the interplay of teams of chemists, biologists, biochemists and physiologists is vital. Additionally, input from toxicologists and environmental scientists is extremely important to the research process. Techniques based upon molecular biology are important in underpinning basic research, especially by facilitating the production of relatively large amounts of target proteins and by enabling the validation of targets. Nevertheless, the biological screen remains a key component since it can be constructed to reflect present and future markets, and thereby act as a representative navigation aid to the discovery effort. In broad terms, there are four conceptual origins which can be taken as starting points for research:

5.1 Random Leads

This method relies upon the biological screen to generate leads. Accordingly, chemicals obtained at random (either by synthesis, from collaborators or from other parts of the chemical industry) are screened and the positive hits which emerge are pursued. The biological screen is carefully constructed to reflect commercial targets, present and predicted. This approach is often referred to as random synthesis or "blue sky" chemistry. It should be noted, however, that whereas the generation of the first hint of activity is indeed random, the process by which the lead is established and optimised to provide a compound with commercial levels of activity is very highly skilled. It will involve a co-operative application of expertise in many areas ranging from, for example, computer-aided molecular modelling at the inception stage through *in vitro* biochemical screens to whole cell and physiology studies.

5.2 Analogue Chemistry

The second method is to take a lead from a competitor's area of activity. Such information is available in the patent literature. The objective then is to develop the lead into a novel and patent-free area, followed by structural optimisation to provide the desired product. While this is the most reliable way to achieve biological activity, there exists the possibility of insufficient differentiation of products so derived *ie* "me-too" molecules. Furthermore, the derivative products will normally share the same mode of action as the parents.

5.3 Natural Products

Clues can be taken from chemicals which occur in nature. Mention has already been made of pyrethroids as an example of outstanding success in this class. Unfortunately, there are relatively few other notable cases under this heading. Aside from the inherent chemical complexity which characterises many natural products, the chemicals which occur in nature frequently have insufficient potency or have the wrong specificity to act as leads for synthesis.

In recent times however, another major area has emerged in the fungicide sector viz the ß-alkoxyacrylates (see 6.3 below).

5.4 Biorational Design

The most intellectually attractive concept is that of biorational design. This requires consideration of the plant, fungus or insect as a biochemical machine. A target enzyme (or receptor) is chosen for attack based upon criteria such as physiological importance and level of occurrence. It is of course vital that inhibition of this enzyme is fatal to the organism concerned. Inhibitors for the target enzyme are then rationally designed and synthesised. Large amounts of certain enzymes have been made readily available in recent times by the application of molecular biology techniques. It is now possible to produce sufficient enzyme to allow crystallisation and subsequent X-ray analysis. This can lead to construction of a molecular model (by computer-aided technology) against which to design inhibitors. Although this approach has enjoyed only limited success to date, it is clear that the area is advancing rapidly. However, rational design of inhibitors active *in vitro* is only part of the story. Candidate inhibitors must also be successful in uptake, translocation and survival of metabolism within the plant or organism.

6 CASE HISTORIES OF SOME IMPORTANT EXAMPLES

6.1 Random Screening - The Discovery of Glyphosate

An example of the random screening approach is provided by the discovery of glyphosate, probably the greatest commercial success in the industry to date.

Work at Monsanto before 1970 led to the synthesis of over 100 aminomethyl phosphonic acids, aimed at non-agricultural screens[5]. However, subsequent biological evaluation of this group using agrochemical screens showed two compounds with interesting activity as plant growth retardants. There was also some interesting herbicidal activity against perennial species, although this activity was too low for commercialisation.

It is interesting to note that compound (17) was active as a sugar cane ripener and was later marketed. Many more analogues were made in an attempt to optimise herbicidal activity, but progress was very limited until the activity

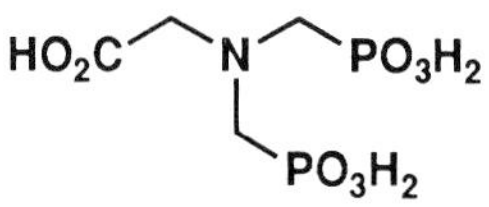

HO_2C–N(PO_3H_2)–PO_3H_2 (17)

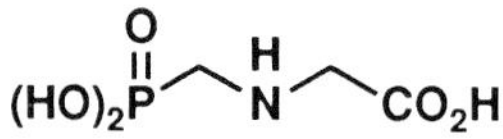

$(HO)_2P(O)$–N(H)–CO_2H (18)

of glyphosate (18) was unearthed. The interesting biological property of this molecule is that its activity is limited entirely to post-emergence treatments *ie* when sprayed onto the foliage of growing plants. Glyphosate is totally inactivated on contact with soil and consequently has almost zero residual effect. This, taken together with benign toxicology and a favourable variable product cost, has led to glyphosate becoming a most remarkable success worldwide.

The mode of action of glyphosate has been shown to be competitive inhibition of 5-enolpyruvylshikimate-3-phosphate (EPSP) synthase. This is an important enzyme in the shikimate pathway, which leads to the biosynthesis of aromatic amino acids and numerous phenylpropanoid biosynthetic intermediates.

6.2 Analogue Chemistry - Aryloxyphenoxypropionates[6]

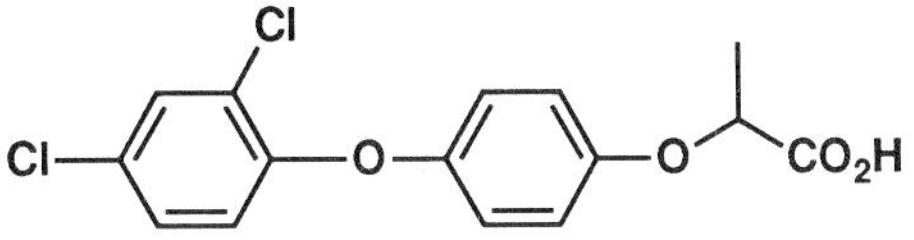

Whereas the commercial success of several aryloxyphenoxypropionate herbicides (AOPs) is an excellent example of the analogue chemistry approach, it is interesting to note that this series of herbicide grasskillers has its roots in the pharmaceutical industry. Hoechst are credited with the invention of the AOPs. They filed two patents to phenoxyalkane carboxylic acids in 1969 and 1971 as pharmaceutical preparations (hypolipodemic agents). In 1973, Hoechst filed UK patent 1423006 claiming herbicidal activity for this type of compound. The patent did not have the same chemical scope as the earlier pharmaceutical claims, but included some specific compounds claimed earlier. Hoechst developed three such compounds as herbicides based upon three acids, given the trivial names diclofop (18), clofop (19), and trifop (20). The first named compounds

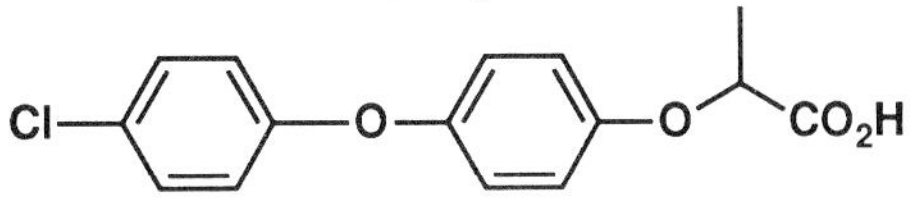

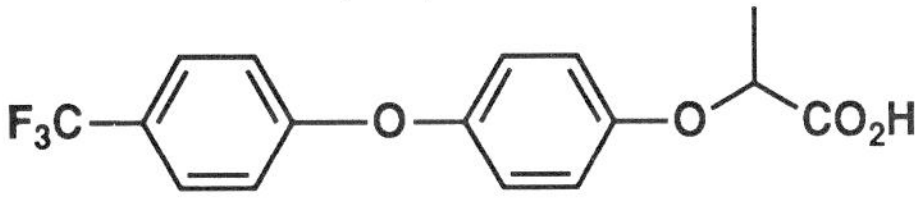

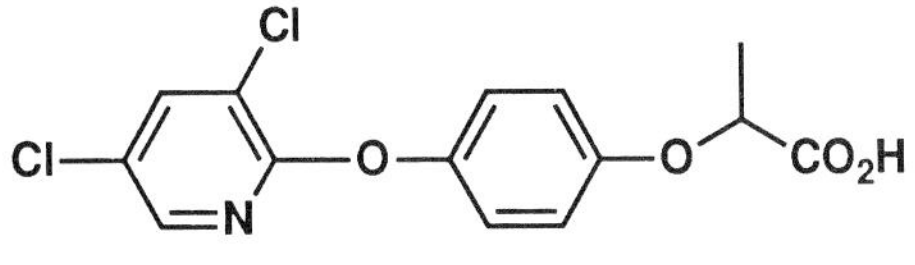

(18)

(19)

(20)

(21)

are cereal selective grass killers, whereas trifop is a wider spectrum grass killer for use in broad leaved crops. This latter compound was not covered by UK 1423006 and was only developed after diclofop and clofop. The AOP compounds are invariably applied as esters, but it is the corresponding acid which is the actual active compound. Of the three compounds mentioned above, only diclofop methyl was eventually commercialised.

Ishihara (ISK) made an important further advance by substituting a pyridine moiety for the benzene ring of diclofop. They developed chlorazifop (21) which has similar herbicidal properties to trifop. Just over a year later, ISK, ICI and Dow filed on the para-trifluoromethyl compounds within three weeks of each other. The ISK/ICI compound was commercialised as fluazifop butyl (22), and Dow introduced haloxyfop methyl (23).

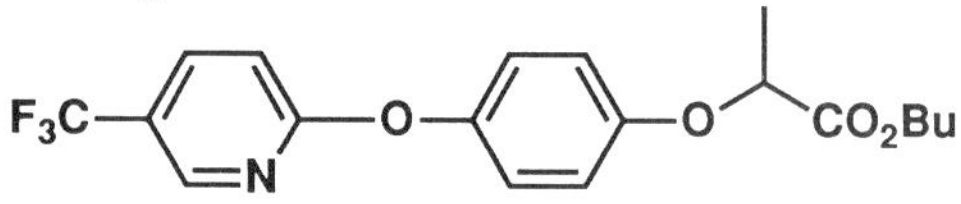

(22)

(23)

The propionate moiety presents a source of chirality in these molecules. Subsequently, ICI filed patents on the single enantiomers of fluazifop and haloxyfop. Of particular interest here is some unexpected biology. In post-emergence tests, the racemate was approximately half as active as the pure R-enantiomer, the S-isomer being almost inactive. In pre-emergence tests however, both are equiactive. In due course, this was shown to be due to rapid microbial conversion of the inactive S-acid to the R-acid in the soil.

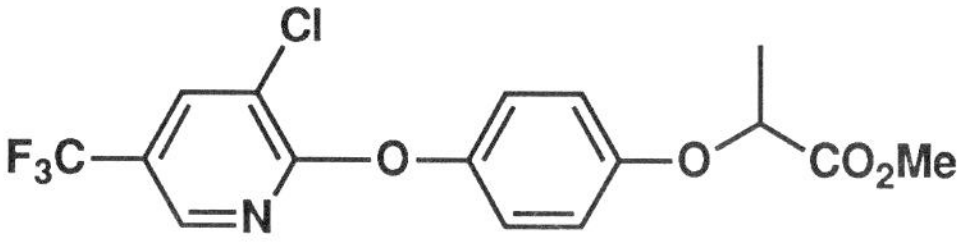

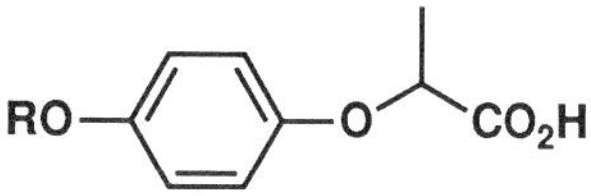

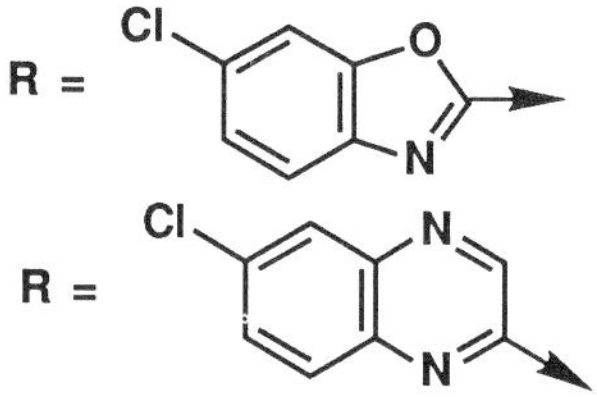

(24)

(25)

Hoechst finally re-entered the fray with fenoxaprop (24) which is moderately cereal selective (normally requires a safener). Other companies also discovered useful bicyclic ring replacements such as the independent discovery of quizalofop (25) by Nissan, DuPont and ICI - a patent race in which ICI came a close third! Many other ring systems are possible at this position with retention of useful activity.

The mode of action of the AOPs is fatty acid biosynthesis inhibition *via* acetyl-CoA carboxylase inhibition. It is of interest that the crop selectivity observed for the series is expressed at the enzyme level as, for example, haloxyfop inhibits the maize enzyme but not that derived from peas. Thus, the compounds enjoy their major usage as grasskillers with extremely good margins of safety in broad-leaved crops.

6.3 Natural Products - ß-Alkoxyacrylates[7]

Methoxyacrylate fungicides derive from a family of natural products isolated in the 1960's and 1970's. The strobilurins and oudemansins are found in several genera of small agarics, including species of *Strobilurus* and *Oudemansiella*, which grow on decaying wood. The ability of these species to synthesise fungicidal compounds

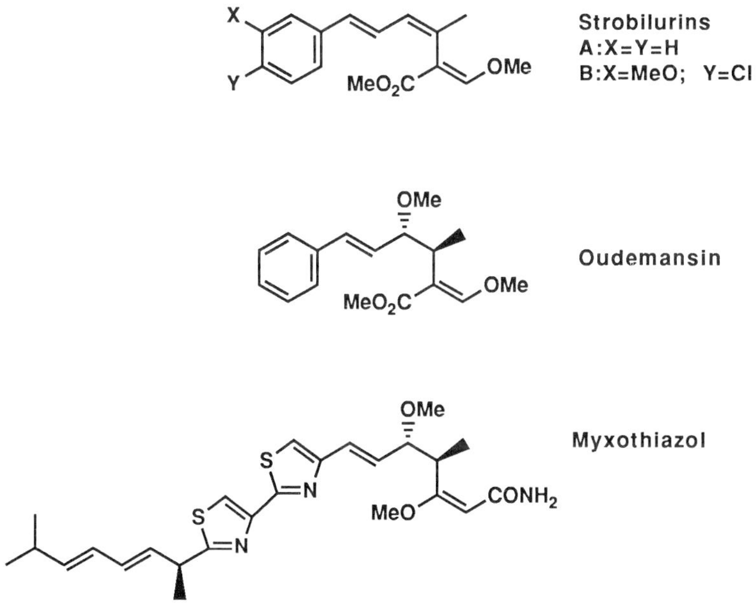

Strobilurins
A:X=Y=H
B:X=MeO; Y=Cl

Oudemansin

Myxothiazol

Figure 3 - Methoxyacrylates

presumably gives them an advantage as they compete for nutrients with other fungi in their natural environment.

In 1981, Anke and Steglich *et al* showed that four compounds of this type (Figure 3) shared the same biochemical mode of action viz inhibition of mitochondrial respiration by blocking electron transfer between cytochrome b and cytochrome c_1. It later became clear that all four compounds bind at the same site on cytochrome b. This is a novel mode of action and hence no cross resistance to known fungicides would be expected.

This natural product lead provided an attractive starting point for synthesis. Initially, both ICI and BASF were involved, the latter in collaboration with Professor Steglich. Oudemansin was tested by ICI and shown to have a broad spectrum of fungicidal activity at approximately 25 parts per million. Strobilurin A was synthesised by ICI and was shown to be a different geometric isomer to that reported in the literature. The activity claimed for the compound was confirmed in *in vitro* tests, but *in vivo* no glasshouse activity was obtained. This was presumed to be due to a combination of photochemical instability and volatility. Accordingly, an analogue programme was commenced with the objective of overcoming these problems and increasing activity.

A stilbene analogue (26) was prepared with the intention of removing the conjugated diene and holding the molecule in the presumed correct conformation. This stilbene analogue was shown to be active *in vivo* but still insufficiently photochemically stable for further progression. Further analogue synthesis was thus required (guided by biological tests, physicochemical parameters and computer graphics etc) and in due course led to several commercially interesting analogues, showing broad spectrum activity against a range of commercially important diseases.

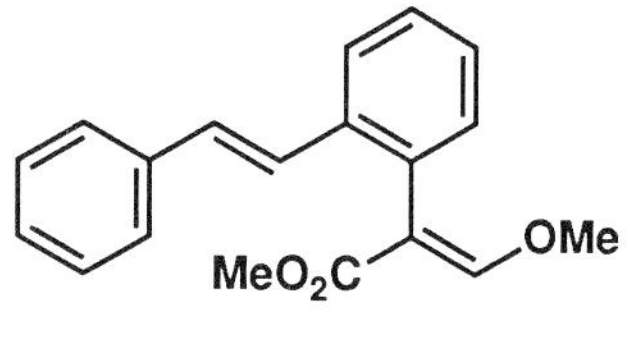

OMe
MeO_2C

(26)

Whereas this area of endeavour emanated from a natural product lead, there have been some 75 patents from 10 companies published to date, testifying to the importance of the area. It is also clear that what originated as a natural product lead has now moved extensively into an area of analogue exploitation. It will be most interesting to observe which areas of chemistry will lead to commercial products in the future - there are none to date.

6.4 Rational Design - Inhibitors of Pyruvate Dehydrogenase

The rational design method has so far failed to provide a commercial product based upon the *de novo* design of an inhibitor of an enzyme for which no agrochemical importance was previously established. However, there have been several near misses reported to date. Whereas the design of potent inhibitors meets with success relatively frequently, there are additional molecular requirements for uptake, translocation, survival of metabolism, toxicological and environmental safety, and indeed the ability to compete with pre-existing compounds in the market place. It is clear that the properties required for potent enzyme inhibition are very unlikely to be those which are optimal for the other processes

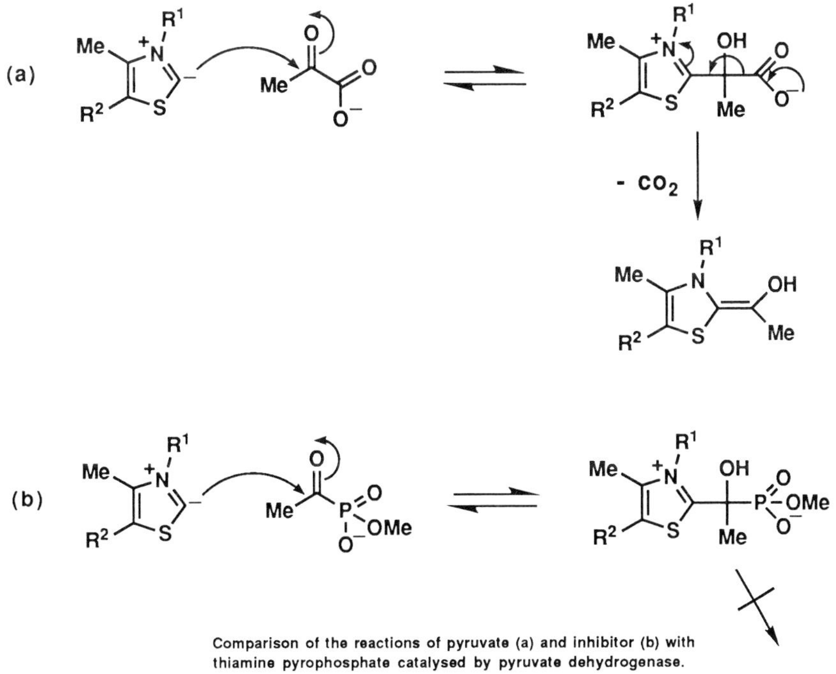

(a) Me R[1] N R[2] S Me O O O⁻ ⇌ R[1] Me N OH O R[2] S Me O⁻ - CO_2 R[1] Me N OH R[2] S Me

(b) R[1] Me N R[2] S O Me P O OMe O⁻ ⇌ R[1] Me N OH O R[2] S Me P OMe O⁻

Comparison of the reactions of pyruvate (a) and inhibitor (b) with thiamine pyrophosphate catalysed by pyruvate dehydrogenase.

Scheme 1

involved. Accordingly, a trade-off in properties is inevitably necessary.

An excellent example of an attempt to design an agrochemical using biochemical reasoning is provided by the invention of inhibitors of pyruvate dehydrogenase as herbicides[8]. Whereas activity was demonstrated at an impressive level in the field, the development was terminated due to insufficient commercial potential.

The mechanism of pyruvate dehydrogenase requires the addition of deprotonated thiamine pyrophosphate to the keto-carbonyl of pyruvate, followed by decarboxylation. This is shown in scheme 1. It was recognised that an acyl phosphonate might act as an effective inhibitor because addition to the carbonyl carbon should take place readily to provide an intermediate which should be strongly bound but unable to complete the decarboxylation step. Series of acyl phosphinates and phosphonates were prepared as mechanism-based inhibitors and indeed some of these were very powerful inhibitors and also herbicidal. A substantial body of evidence was obtained through biochemical testing to support strongly the view that susceptible plants died as a direct result of the enzyme

Table 3 - Structures of Compounds and Enzyme Inhibition Data

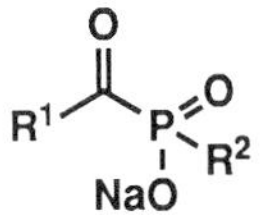

O
R^1 P=O R^2
NaO

Compound	R^1	R^2	Inhibition data
I	CH_3	OCH_3	I_{50} = 70 μM[a]
II	CH_3	CH_3	$t_{1/2}$ = 12 min at 10 μM[b]
III	CH_3	H	$t_{1/2}$ = 3 min at 0.3 μM[b]
IV	CH_3	C_2H_5	I_{50} = 165 μM[a]
V	H	CH_3	Inactive
VI	C_2H_5	CH_3	$t_{1/2}$ = 12 min at 1 mM[b]

[a] Reversible.
[b] Time-dependent

inhibition. Some examples are provided in table 3 together with inhibition data. As stated earlier, none of the compounds had sufficient commercial potential for full development, but the results of the study were sufficiently promising to encourage the view that biochemical design of agrochemicals will provide success in the future.

7 THE FUTURE

Against the backcloth of a world in which the population is expected to double in the next 40 or 50 years, in which the requirement for better quality and more variety in foodstuffs will become more widespread and in which there are essentially no competing technologies, it is easy to be optimistic about the future for crop protection chemicals.

However, the characteristics required for new products will change significantly. In order to recompense the rapidly increasing costs of research and development, industry will be obliged to develop high added value chemicals. Increased emphasis will need to be placed upon several of the following criteria for new products:

- very highly active and cost effective
- even safer to the environment, the user and the consumer
- compatible with other products and suitable for use in integrated pest management programmes
- flexible in use

Compounds with novel modes of action will be highly prized in several areas, since they will be useful in strategies to avoid resistance. The physico-chemical properties of molecules will be important especially if they enable efficacious formulations.

With regard to the impact of biotechnology, rapid advances will be made. This will lead to the introduction of novel crop varieties and microbial products in the next few years. Some of these introductions will have considerable local importance, but it it unlikely that they will take more than a 5% share of the present total crop protection market by the year 2000. The rate of commercial introductions is likely to be constrained not only by economic and technical factors, but also by the regulatory framework which is being established in many countries to govern their development.

Perhaps the major task facing the industry in the near future is to reverse the negative public perception of the chemical products of the industry. Undoubtedly, the industry deserves some of the bad press that it receives because in the past it has failed to take steps to effectively counter the sometimes misplaced and exaggerated claims of pressure groups. The industry can be expected to take positive steps to effect a change in this negative perception. Information on crop protection chemicals and their benefits will be much more widely available both to the public at large and to groups such as schools, universities and learned societies. Rather than enter confrontation with environmental pressure groups, the industry will seek closer involvement and collaboration with such bodies. The industry will seek to show that its massive investment in regulatory and environmental work is capable of producing a scientific database which can assure the public of safety. There will be a marked increase in safety, health and environment activity and awareness in the companies, and this will be paralleled by increases in capital expenditure to achieve improvements in operations. At the farmer level, the companies will pay ever more attention to training and product stewardship, and will ensure that the standards which are adopted in the developed world will also be applied universally to developing countries.

From the foregoing discussion, it will be clear that the future for crop protection chemistry research and development is one which will become increasingly rooted in its science base. This science will be increasingly multi-disciplinary, but one in which the chemical sciences will continue to play a major role. It is therefore most appropriate that as we close down the first century of crop protection chemicals, that at least the early part of the next century presents a welter of interesting challenges to the chemical community.

8 ACKNOWLEDGEMENTS

The authors wish to acknowledge Mrs I Dowling and Mrs L A Hall for secretarial assistance and Dr J R Hadfield and Mrs S L MacPherson for helping to prepare visual aids and graphics.

REFERENCES

1 Agrochemical Monitor, County Natwest Woodmac p5, No 74, 22 February, 1991.
2 *Ibid*, p3.
3 Data taken from B.G. Lever, "Crop Protection Chemicals", Ellis Horwood (UK), 1990, p59.
4 For a more comprehensive review, see B. G. Lever, *ibid*, Chapter 3, pp 44-80.
5 J.E. Franz, "The Herbicide Glyphosate", ed. E. Grossbard and D. Atkinson, Butterworths, London, 1985, Chapter 1.
6 K. Beautement, J.M. Clough, P.J. de Fraine and C.R.A. Godfrey, *Pestic. Sci.*, in press.
7 D. Cartwright and J.W. Dicks, personal communications.
8 A.C. Baillie, K. Wright, B.J. Wright and C.G. Earnshaw, *Pestic. Biochem. Physiol.*, 1988, *30*, 103.

The Contribution of Chemistry in Animal Health

R.J. Willis

SCHERING AGROCHEMICALS LTD., CHESTERFORD PARK RESEARCH STATION, SAFFRON WALDEN, ESSEX CB10 1XL, UK

1 INTRODUCTION

The world animal health and nutrition products market in 1988 was estimated to be $9,365 million at the manufacturer's level and $11,040 million at the end user level.[1] The main sectors are biological products (12.8%), medicinal feed additives, such as antibacterials (17.2%), nutritional feed additives (28.5%) and pharmaceuticals (41.5%). The latter sector includes chemicals for the control of external parasites (ectoparasites) and internal parasites (endoparasites), with combined market value of $1,542 million, and it is the aim of this article to review the significant advances made over the last 150 years in this area. The chemical control of parasitic diseases[2] and arthropod parasites[3] of livestock have been comprehensively reviewed elsewhere.

It is also important to recognise here the wider contribution of the pharmaceutical industry, for example in the development of vaccines for control of devastating animal diseases such as rinderpest and foot and mouth, and in the development of antimicrobials, used both as feed additives and as therapeutics. A further significant achievement has been in the development of antiprotozoals such as the coccidiostats, which have had a major impact in the expansion of the poultry industry worldwide.[4]

2 PARASITES AND ECONOMIC LOSSES

Examples of the main helminth endoparasites and ectoparasites which infest livestock are indicated in Table 1. These main types are geographically widespread, although the economic importance of each in a particular area depends on such factors as climate and farming practices. In the UK alone, there are some 20 species of gut nematodes and 4 species of lungworms which infest sheep,[5] and in North America[6] 22 species of ticks, 7 species of mange mites and 10 species of flies are known to infest cattle.

Economic losses are difficult to estimate because of the lack of information from controlled studies.[6] Historically, diseases such as rinderpest have probably been more obviously devastating than losses through parasitisation.[8] However, in 1879, one of the wettest summers on record, the liver fluke may have been responsible for the loss of up to 3 million sheep in the UK.[7] Often the effects of parasites on production may not be so obvious, with no clearly defined deterioration in the animal, but productive grazing may be affected by discomfort before serious losses occur.[6]

A further complication arises, for example in Africa, where parasites such as cattle ticks are important vectors of diseases such as babesiosis, theileriosis and anaplasmosis.[2] It is important that all life stages of the parasite invading host animals are rapidly controlled to prevent disease transmission.

Recent estimates suggest that, even in high income countries with the ready availability of effective remedies, cattle losses through mortality and morbidity (all diseases and parasites) may be 5% and 10% of output respectively[9] (£600 million total in the UK). Losses in developing countries may be at least twice as high as a percentage of output, which is probably a better indication of the scale of the problem that existed more generally in the 19th century.

3 HISTORICAL REVIEW 1841-1941

Pharmacology

Key factors in the development of pharmacology in the 19th century were the expanding knowledge of chemistry, which led to the isolation and identification of the active ingredients in ancient herbal remedies, [10,11] e.g.

	Species	Damaging effects
ACARINES		
Cattle ticks	*Boophilus spp* *Rhipicephalus spp* *Amblyomma spp*	Blood loss, paralysis, hide damage, secondary infection, disease vectors
Mange mites	*Chorioptes spp* *Psoroptes spp* *Sarcoptes spp*	Scab formation, irritation, hide damage, secondary infections, loss of fleece or hair
INSECTS		
Myiasis flies	*Lucilia spp*	Tissue, fleece and hide
	Cochliomyia spp *Hypoderma spp*	damage, toxicosis, secondary infection
Biting and sucking flies	*Haematobia spp* *Stomoxys spp*	Blood loss, irritation disease vectors
Face flies	*Musca spp*	Irritation, disturbance of feeding
NEMATODES		
Gut worms	*Haemonchus spp*	Anaemia
	Ostertagia spp *Trichostrongylus spp* *Cooperia spp* *Nematodirus spp*	Gastro-enteritis, tissue damage, depressed growth
Lung worms	*Dictyocaulus spp*	Parasitic bronchitis
CESTODES		
Tapeworms	*Moniezia spp*	Tissue damage
TREMATODES		
Liver flukes	*Fasciola spp*	Anaemia, weight loss, death

Table 1 Some examples of livestock parasites of economic importance

quinine in 1820, and the realisation that a relationship exists between the chemical composition of a substance and its physiological action. Paul Ehrlich[12] (1854-1915), who is widely acknowledged as the founder of modern chemotherapy, appreciated the need for a selective action on the parasite relative to the host animal.

By the mid 19th century *in vitro* experiments with test substances had been carried out on freshly taken parasitic worms, and natural infestations in cats and dogs had been used to test especially active materials *in vivo*.[11] However, it was nearly a century later before use was made of experimentally induced infections in small animal models for large scale screening of potential anti-parasitic chemicals.

Typical Remedies

Herbal. These usually contained a wide range of readily available ingredients, some of quite long standing use, which were nevertheless of dubious value in the treatment of parasites.[11]

Two widely used anthelmintics for man and animals were wormwood and oil of chenopodium. Medicinal wormwood, obtained from the dried unexpanded flower heads of *Artemisia cina* (Levant wormseed) contains the sesquiterpene lactone santonin (1) and an essential oil. More recent publications suggest that santonin is a nerve poison, which is effective only at doses that may be toxic.[13] Oil of chenopodium or American wormseed, obtained from the leaves, seeds, flowers and roots of *Chenopodium ambrosoides* contains the active principal ascaridole (2), which is also toxic[13] and may do little more than suppress worm egg production.[11]

The inclusion of tobacco leaves in the preparation of insecticidal treatments was also quite common, and the active principal, nicotine (3), was later available as a 40-50% solution of the sulphate. Pyrethrum powder obtained from pyrethrum flowers *Chrysanthemum cinerariaefolium*, now known to contain pyrethrins I (4) and II (5), as key active components, was in widespread use for control of lice and ticks on sheep by the turn of the century.[2]

A further significant development in the early part of the 20th century was the introduction from the Far East of powders and extracts of derris root, later known to include the insecticide rotenone (6). This was of particular value in the treatment of warbles[2] when applied as an ointment, wash or dust to the backs of cattle. It was still quite widely used up to the 1950s in the USA after being temporarily unavailable during World War II.

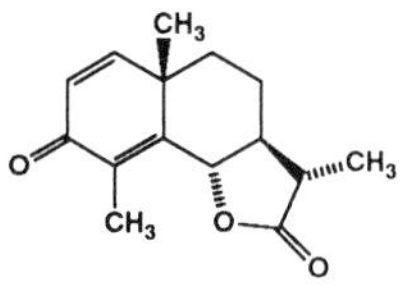

(1) 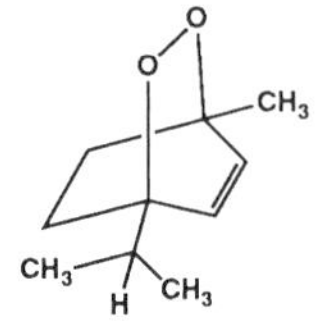 (2)

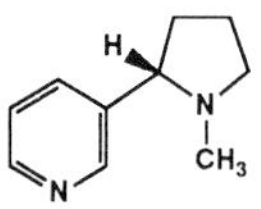

(3)

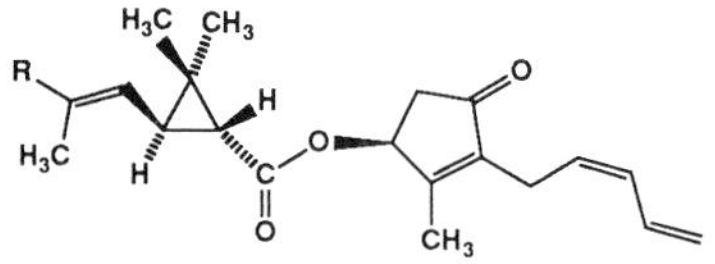

(4) R = CH_3
(5) R = CO_2CH_3

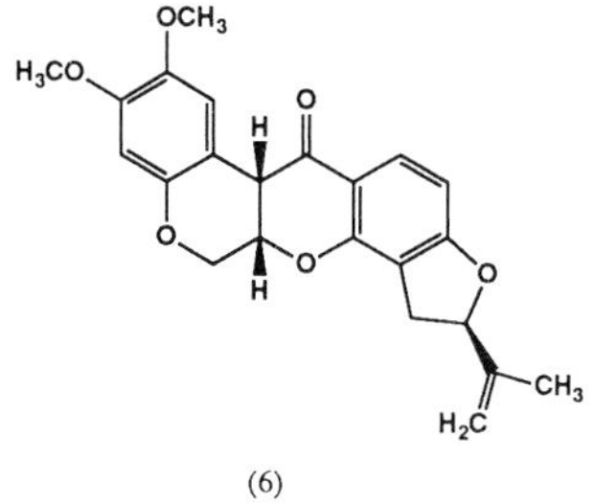

(6)

Figure 1 Active components in herbal remedies

Non-specific Poisons. Mercury compounds had been used for de-worming sheep from the 16th century[11] and mercury dressings were widely used during the 19th century for control of blowfly problems in England.[14]

Arsenic preparations such as arsenic trioxide, often in combination with sulphur, were more favoured for dipping sheep and cattle and were used well into the 20th century.[14] Arsenic was extremely successful in the USA in the cattle tick eradication campaign in the early 20th century. It was also the first practical method of tick control in Australia, South Africa and South America when it was introduced around the turn of the century.[15]

Copper sulphate, with or without nicotine or arsenic, was widely used for worming sheep right up to the 1950s.[16]

Other Materials. Various natural oils, such as oil of tansy, castor oil and cotton seed oil were used either alone or in admixture with some of the above ingredients. Coal tar and petroleum products such as kerosene emulsions were also used and in the late 19th century phenol based dips were introduced for safer control of sheep scab.

Carbon tetrachloride and hexachloroethane were introduced as the first effective treatments for liver fluke in the 1920s although there were toxicity problems.

Development of the Industry

During the 19th century, there was an expanding trade in veterinary medicines supplied by dispensing chemists under their own brand names to recipes obtained from farmers, farriers or veterinary surgeons.[17] Ingredients were mainly cheap imports and many of the concoctions offered little improvement over the old herbal remedies.

However, some reputable companies did develop during this period and it is worth briefly noting the origins of Coopers,[18] which is a widely recognised name in animal health. The founder William Cooper was a veterinary practitioner who moved to set up practice in Berkhamsted in 1843.[19] He developed a sheep dip based on arsenic and sulphur for the treatment of sheep scab, which was the major parasite problem in England at the time, and his first factory opened in 1852. Apparently he was a careful formulator who diligently recorded details of his experiments in notebooks, some of which survive. Other remedies were developed, including fly powders and phenolic sheep dips, and the business was expanded overseas. Chemistry and entomology laboratories were established at Berkhamsted in 1907.

Major rivals were McDougall Brothers who had also started producing sheep dips in the 1850s and were the first to commercialise the use of derris root early in the 20th century. In the 1920s these rivals combined with another long established firm to trade as Cooper, McDougall and Robertson and they continued to build on their expertise as formulators of anti-parasitic products. In modern times the business developed under the ownership of The Wellcome Foundation (1959), Wellcome and ICI jointly (1984) and Pitman-More of the American IMC Group (since 1989), which is now one of the world's largest animal health companies.[20]

4 THE MODERN ERA

As in many areas of applied science, progress in animal health has been spectacular since the Second World War. The successful wartime development of e.g. antibiotics in the pharmaceutical area and DDT in the agrochemical/ public health sector paved the way for the expansion of research. The increasing sophistication of chemistry and the biosciences has given rise to a variety of effective synthetic molecules.

Ectoparasiticides

While the natural insecticide pyrethrum was reasonably effective, it was quite unstable in light and therefore of short persistence. The search for more effective alternatives was also stimulated by the interruption in the supply of derris during the war.

Organochlorines. In 1939 the Swiss firm Geigy discovered that dichlorodiphenyltrichloroethane (DDT) (7) has a powerful and persistent insecticidal action. When it became available commercially for tick control in 1946 it was regarded as the first safe and effective treatment.[15] It achieved wide use on animals and their housing.[17,21] The story of DDT is too well known[22] to require further detailing here, suffice to say that it was withdrawn from main markets by the early 1960s because of unacceptable residue levels in meat.

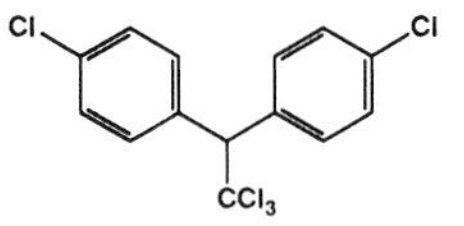

(7) 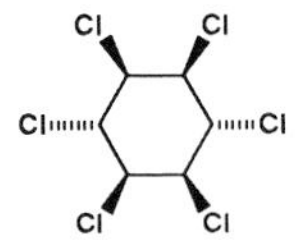 (8)

Figure 2 Organochlorine insecticides

Other organochlorine materials such as gamma-HCH (8), dieldrin and campheclor,[23] which have a different mechanism of action, also achieved significant use. These compounds continued to play a role in ectoparasite control well beyond the withdrawal of DDT, e.g. camphechlor for control of ticks.

Organophosphorus Compounds. Following the pioneering work of Bayer in the 1940s, a number of organophosphorus compounds became available for ectoparasite control before the withdrawal of DDT, e.g. dioxathion (9) (1958) and coumaphos (10) (1959).[15] This group of compounds achieved widespread even dominant use and continue to be very effective as commodity products, although resistance now precludes their use in some markets. It is interesting to note here significant advances in methods of application with some of these compounds. The control of myasis producing larvae such as ox warbles or cattle grubs (*Hypoderma spp*) using rotenone was not completely satisfactory, which stimulated the search for insecticides which might act systemically, i.e. after oral administration,[23] to kill the larvae migrating through the body of the animal. Ronnel (11) was the first practical systemic insecticide to control cattle grub larvae before they reached the backs of cattle thus reducing hide damage and loss of meat through trimming at slaughter.[2]

Other compounds were subsequently shown to be systemic by spray application and/or by the pour-on technique, in which a small volume of insecticide concentrate is poured along the backline of cattle. Some organophosphorus compounds also had activity against internal parasites but the availability of safer alternatives limited their success.

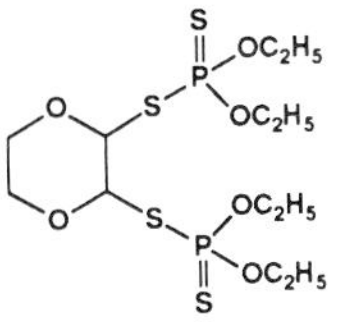

(9)

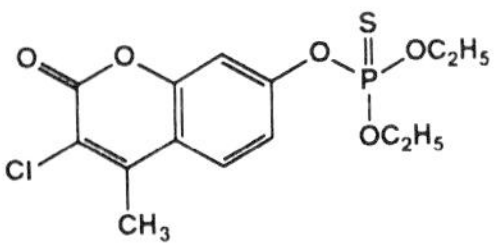

(10)

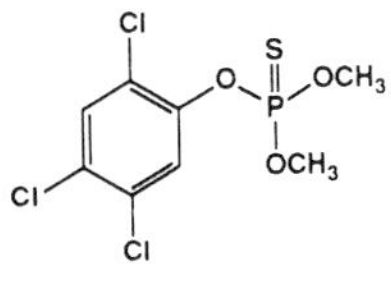

(11)

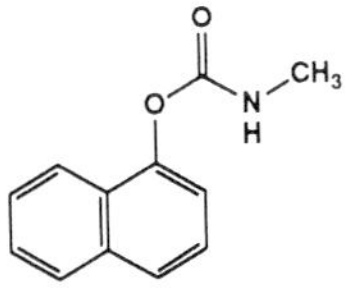

(12)

Figure 3 Organophosphorus and carbamate ectoparasiticides

Carbamates, which have the same mode of action as the organophosphorus compounds,[24] did not generally achieve such widespread use because of less favourable toxicology. However, carbaryl (12) has been used to control flies, mites and ticks with adequate safety.[3]

Amidines. This smaller class of compounds is typified by amitraz (13),[25] which is active against mites and ticks. This compound has a distinct rapid expellent action, which is of particular value where ticks are disease vectors. It was commercialised in the 1970s when severe resistance to organophosphorus compounds was developing in Australia,[15] and it is still one of the few compounds able to control the newer strains of pyrethroid resistant ticks.

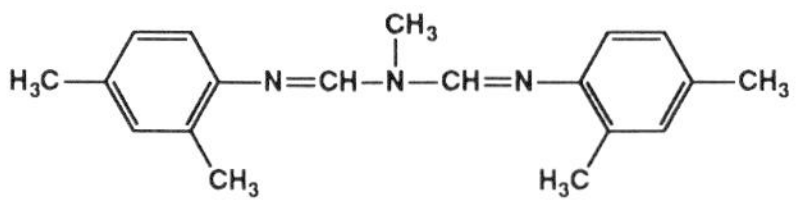

(13)

Figure 4 Amitraz, an ectoparasiticide with novel effects

Pyrethroids. The natural pyrethrins mentioned earlier gave rise to a vast area of industrial research, largely inspired by the work of Michael Elliott and co-workers at Rothamsted Experimental Station,[26] which has led to a wide range of synthetic pyrethroids of greatly enhanced efficacy and stability for both crop protection and animal health. Careful scientific study over many years at Rothamsted established the key features required for activity including the most appropriate stereochemistry at the asymmetric carbons in both natural and synthetic analogues. This culminated in the discovery of deltamethrin (14), one of the most potent known insecticides, which is a single diasteroisomer. One of the major achievements of the agrochemical industry has been to produce such compounds on a industrial scale at an economic price. Of particular note is that deltamethrin crystallises from a mixture with its inactive diasteroisomer and that the latter may be completely converted to the former by epimerisation at the benzylic carbon. Deltamethrin and its dichlorovinyl analogue cypermethrin (isomer mixture) have been successfully commercialised for control of ticks and biting flies.

Japanese workers at Sumitomo discovered that the vinyl dimethylcyclopropane ring may be replaced by an α-isopropylbenzyl group to give a series of potent stable insecticides of which fenvalerate (15) and flucythrinate have been commercialised in animal health.[2]

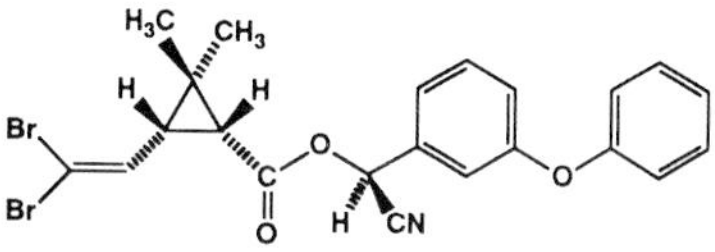

(1*R*) - *cis*, α - (*S*)

(14)

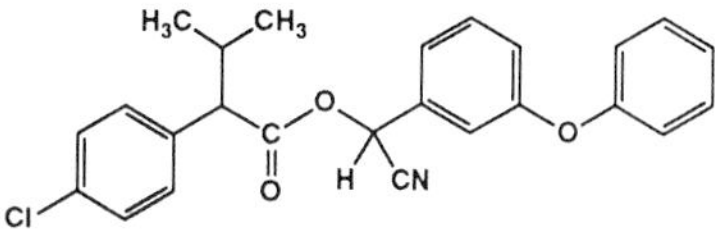

(15)

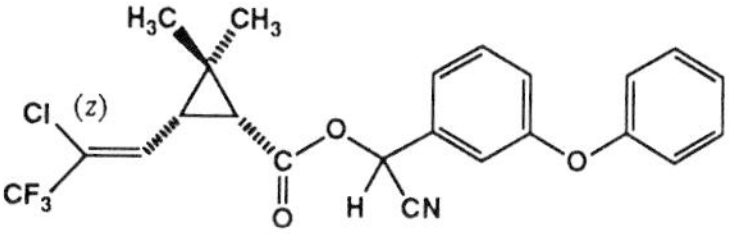

$\geq$ 95% *cis* isomers

(16)

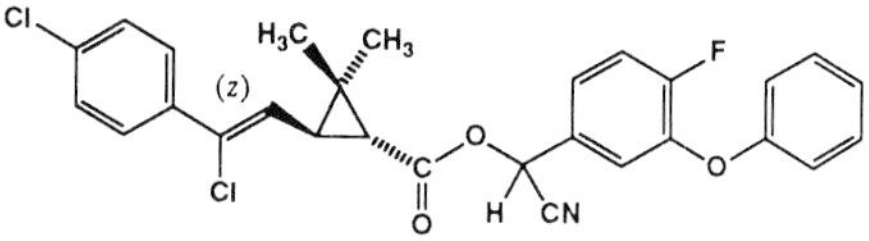

trans racemic

(17)

Figure 5 Some of the key synthetic pyrethroids used for control of ectoparasites

Other compounds of particular note are cyhalothrin (16)[27] and flumethrin (17),[28] which were developed primarily for the animal health market, although cyhalothrin has been subsequently marketed in crop protection. These two compounds introduced in the 1980s are the most active yet known against a variety of susceptible and resistant cattle tick strains and yet have a good safety margin. Flumethrin has been particularly successful as a pour-on treatment, however, after less than five years use in Australia resistant tick strains were identified and the search for new molecules with novel modes of action continues.

Anthelmintics

The major products in this area are for control of the nematode parasites (gut worms and lung worms) of livestock. Other compound classes such as phenols and salicylanilides are effectively used for control of liver fluke and have only limited activity against certain blood sucking nematodes.[2]

Phenothiazine. This synthetic coal tar product (18) has been known since 1885 but its activity against endoparasites was not discovered until 1940. It is therefore worthy of note as the first reasonably broad spectrum worm treatment available as an alternative to the traditional copper sulphate.[21] However, the dose rate of 500-600mg/kg was high by modern standards and the compound was soon superseded by a new class of compounds.

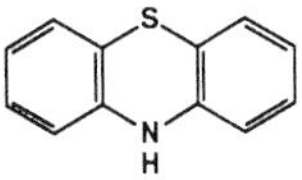

(18) 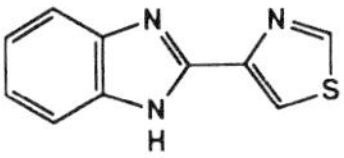 (19)

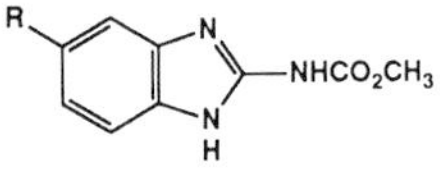

$R = CH_3CH_2CH_2CH_2$— (20)

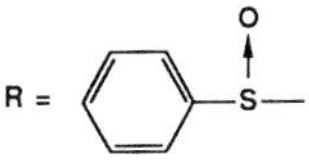

(21)

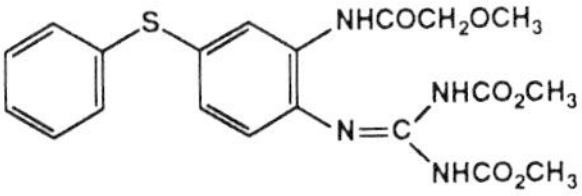

(22)

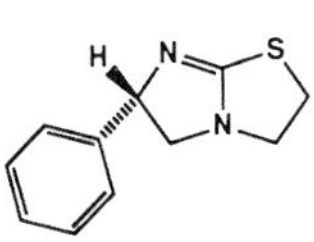

(23) 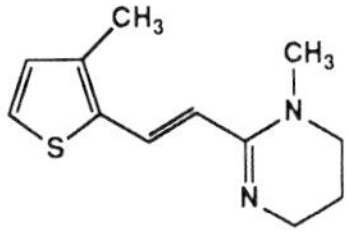 (24)

Figure 6 The post-war development of anthelmintics

Benzimidazoles. The introduction of thiabendazole (19) by Merck in 1961 marked the beginning of a new era of broad spectrum anthelmintics,[29,30] which has given rise to 15 or so commercial drugs and pro-drugs, some of which are still of major significance. It became evident that thiabendazole was susceptible to a rapid enzymic hydroxylation at the 5-position of the benzimidazole nucleus. With the increasing resources of the post-war pharmaceutical industry, large numbers of analogues were synthesised and tested and suitable blocking 5-substituents were discovered to produce drugs with increased half lives.

The major breakthrough came however with the discovery at Smith Kline that the thiazole ring could be replaced by a 2-methylcarbamate substituent. This gave rise initially to parbendazole (20) but other companies soon followed with their own compounds all with the same 2-substituent in combination with other 5-substituents, e.g. fenbendazole (5-phenylthio), oxfendazole (5-phenylsulphinyl) and albendazole (5-propylthio). These compounds offered real advances both in spectrum of activity and lowered dose rates,[30] e.g. thiabendazole is used at 66-110mg/kg in cattle whereas oxfendazole (21) is used at 4.5mg/kg with better control of larval gut nematodes. The compounds are generally safe although concern over teratogenicity has limited their use in pregnant animals and withdrawal periods for meat and milk are variable. A number of pro-drugs, which are converted to benzimidazole carbamates *in vivo*, have also been successfully commercialised, e.g. febantel (22).

<u>Imidazothiazoles and Tetrahydropyrimidines</u>. Two other compound types, which have a similar mode of action,[31] emerged in the 1960s and are still in use today. The first of these is represented by the imidazothiazole, levamisole (23),[32] a broad spectrum compound for control of worms in sheep and cattle, which is active either as a traditional drench or as a pour-on formulation. The compound is safe for use in pregnant animals and has a relatively short withdrawal period. Activity against arrested larval stages and eggs of gut worms is inferior to the most potent of the benzimidazole class.

The tetrahydropyrimidine morantel (24), which is a representative of the second group of compounds, is particularly active against adult stages of gut worms and may be used in pregnant animals. In many areas it is only available as its tartrate salt in a slow release bolus device to give long term protection against worms in cattle.[33]

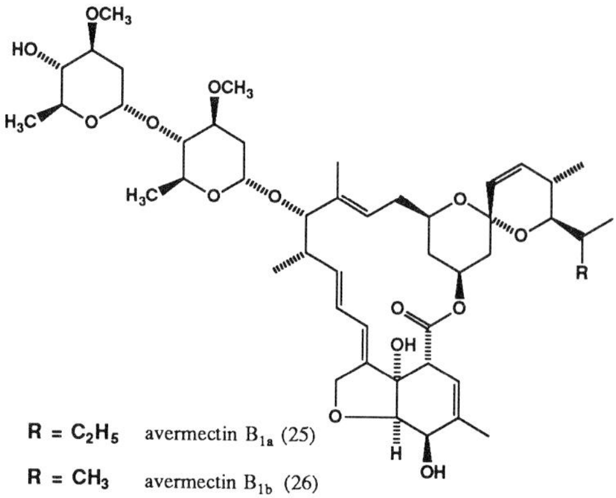

OCH_3 HO H_3C O O OCH_3 H_3C O O O O O R O O OH O H OH

R = C_2H_5 avermectin B_{1a} (25)

R = CH_3 avermectin B_{1b} (26)

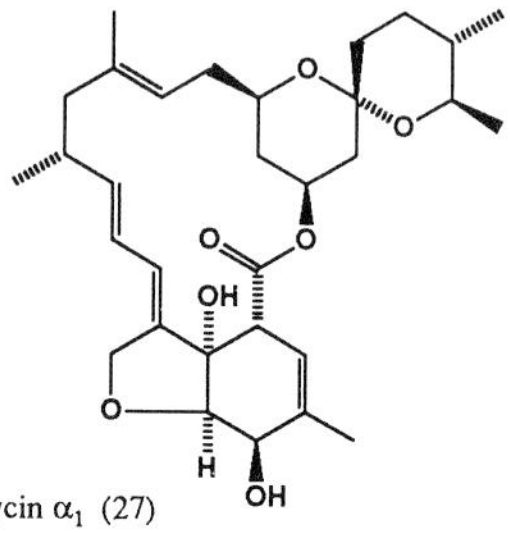

O O O O OH O H OH

milbemycin α_1 (27)

OCH_3 HO H_3C O O OCH_3 H_3C O O O O R O O OH O H OH

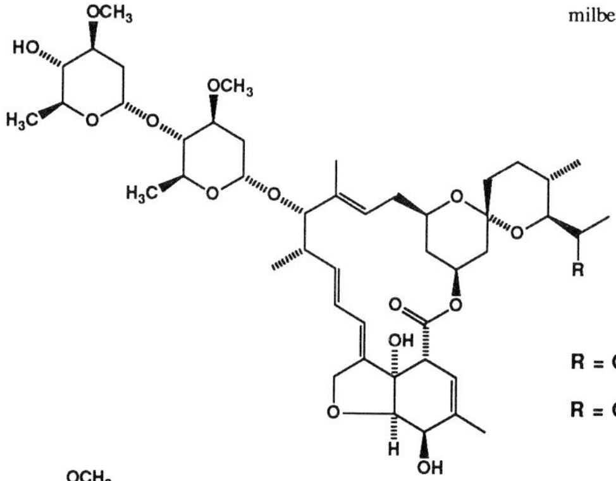

R = C_2H_5 (80%)
R = CH_3 (20%) } ivermectin (28)

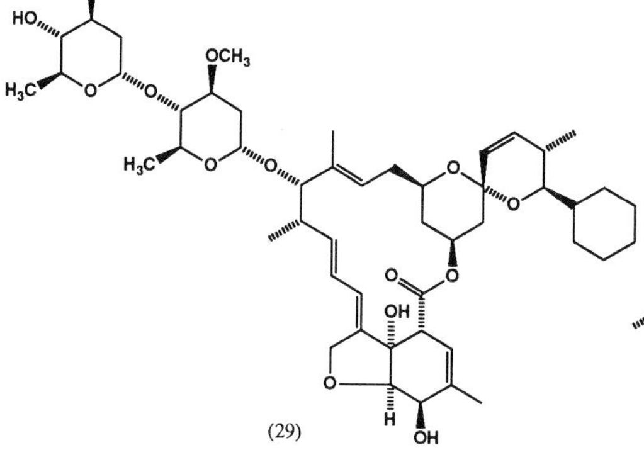

OCH_3 HO H_3C O O OCH_3 H_3C O O O O O O OH O H OH

(29)

CH_3O N O O O O OH O H OH

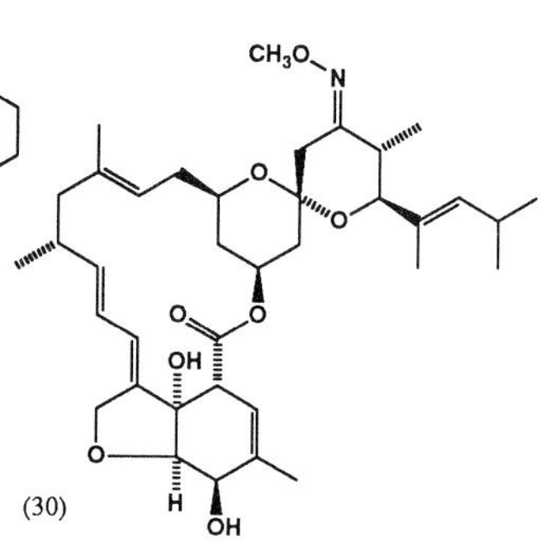

(30)

Figure 7 Avermectin and milbemycin endecticides

Avermectins and Milbemycins. These closely related structure types are both 16-membered cyclic lactones with the main difference being that the avermectins have a disaccharide at the 13-position whereas the milbemycins have no substituents,[34] e.g. compare structures (25), (26) and (27) in Figure 7. The avermectins were discovered by Merck, Sharp & Dohme scientists in a major screening programme on the products of fermentation broths using a specially developed *in vivo* mouse assay. Out of thousands of samples one *Streptomyces* culture, subsequently named *Streptomyces avermitilis*, produced anthelmintic activity, which was soon shown by HPLC analysis to be correlated with the presence of eight related components.[35] The powerful combination of ^{13}C NMR and mass spectrometry enabled elucidation of the complex structures and their relationship to already known milbemycin structures, which was later confirmed by X-ray analysis.

A breakthrough came when Merck scientists discovered that a synthetic dihydro derivative of the component B_{1a} had an improved spectrum of activity. This compound, ivermectin (28), discovered in 1979 has now developed into the largest selling antiparasitic product in animal health. Of particular significance was the discovery that a single compound could control both internal nematodes and external arthropod pests by systemic action and the term endecticide is now used to describe this spectrum of activity.

The effective dose level to control these parasites is 0.2mg/kg by subcutaneous injection, which is significantly lower than other compounds available on the market. Ivermectin is effective against adult and larval stages of nematodes and is safe for use in pregnant animals although, because of its relatively long half-life the withdrawal period for meat is up to 28 days. The compound is active against ectoparasites which include cattle grubs and mange mites but control of cattle ticks is incomplete at normal dose rates.

The great success of ivermectin has stimulated a huge amount of research worldwide. The structures have presented a challenge to academic synthetic chemists and many neuroscientists have studied their mode of action. Other animal health companies have been actively pursuing their own screening programmes on fermentation broths, as is evident from the many published patents on natural products and their synthetic derivatives.

At least one avermectin, doramectin (29) (Pfizer), and one milbemycin, moxidectin (30) (American Cyanamid), are currently under development.

5 DELIVERY SYSTEMS

Traditional methods of application or dosing in animal health include ointments, washes, plunge dips and sprays for control of ectoparasites, and orally administered drenches or pastes for control of helminths. In the last 30 years significant advances have been made in application technology which have brought benefits to the farmer and his stock. The main advantage has been the reduction in time required for handling animals which represents less direct cost to the farmer and reduced stress on the animals. Stress can also affect the weight gain of animals and as such is an indirect cost. Newer techniques include pour-ons, spot-ons, ear tags, collars, injections, bolus devices for slow or pulsed release and implants.

The pour-on technique, which involves the application of a small volume of concentrate along the back of the animal, was first used for systemic control of cattle grubs[23] with organophosphorus compounds in the 1960s and has since been used for control of external or internal parasites and, mostly recently with ivermectin, control of both.

Probably the most important recent advance in delivery has been in the use of controlled slow release boluses.[36] Some of these devices leave metal parts in the rumen but a recent bolus, which consists of a flexible sheet containing the anthelmintic morantel tartrate, is claimed to give 90 days protection after which the sheet gradually disintegrates and is passed out.

6 SOCIAL AND ENVIRONMENTAL ISSUES

Clearly with the newer technology a precise amount of drug is more easily applied to the animal for parasite control thus minimising the impact on the environment. In the 19th century the widespread use of non-specific poisons such as arsenic for dips must have posed a major disposal problem. The compounds developed since the war are more specific in their action and effective doses have been progressively lowered. For example, the effective dose rate to control nematodes in sheep with phenothiazine, thiabendazole, oxfendazole and ivermectin is 600, 44, 5 and 0.2mg/kg respectively.

One of the major changes in livestock farming since the war has been intensification. Increased herd or flock size potentially increases the disease/parasite pressure and such developments would have been impossible without the availability of effective parasiticides, vaccines, antibiotics and feed supplements. Thus the animal health industry has contributed to the increased efficiency which has enabled a steady supply of meat at relatively stable prices.

The issue of intensive farming has been the subject of much debate with pressure on the farming industry to move to more spacious housing and grazing. This has the support of the animal health industry as long as the system of management reduces the incidence of animal and human disease. It is highly doubtful that this can be achieved without the prophylactic use of vaccines, antibiotics and parasiticides.

In the EEC, steroid growth promoters have been banned largely on socio-economic rather than scientific grounds and the routine use of antibiotics also used in human health is under review.

On the positive side, animal parasiticides such as the benzimidazoles[37] and ivermectin have subsequently been developed for applications in human medicine and are of particular value in developing countries.

7 THE FUTURE

The decline in the number of companies active in animal health and nutrition as a result of acquisitions and mergers is likely to continue. One reason for this is that the industry is highly research intensive and the costs for development and product registration are increasing rapidly. The estimated worldwide R&D expenditure is of the order of $850 million.[1]

In the immediate future there is a continuing need for new active molecules with novel mechanisms of action to control resistant parasites.[15,38] Attempts are being made to increase knowledge of the biochemistry and physiology of target parasites to help facilitate this. In the longer term, biotechnology may provide a range of vaccines for use against parasites.

Biotechnology is likely to make a more immediate impact in other sectors of the market with the availability of growth hormones such as bovine somatotropin.

The success of these may well be hampered by socio-economic and political considerations.[39]

REFERENCES

1. "Agrochemicals and Animal Health Services Summary Report", County NatWest Securities Ltd., 1990.
2. W.C. Campbell and R.S. Rew, "Chemotherapy of Parasitic Diseases", Plenum Press, New York, 1986.
3. R.O. Drummond, J.E. George and S.E. Kunz, "Control of Arthropod Pests of Livestock", CRC Press, Boca Raton, 1988.
4. R.B. Griffiths, "Animal Health and Productivity", Proceedings of an International Symposium, Royal Agricultural Society of England, 1985, p.37.
5. H. Hope, Farmers Weekly, 1990, 4 May, p.72.
6. R.O. Drummond, "The Economic Impact of Parasitism in Cattle", Proceedings of the MSD AGVET Symposium, W.H.D. Leaning and J. Guerrero, Veterinary Learning Systems Inc., 1987, p.9.
7. G. Graham, Reference 5, p.85.
8. D. Taylor, Vet.Hist., 1975, (5), 19.
9. T. Wesley, "The Economics of Animal Disease", V&O Publications Ltd., 1987.
10. C.D. Leake, "An Historical Account of Pharmacology to the Twentieth Century", Charles C. Thomas, Springfield, Ill., 1975.
11. W.C. Campbell, Reference 2, Chapter 1, p.3.
12. J. Parascandola, J.Hist.Med.Allied Sci., 1981, 36, 19.
13. O.D. Standen, "Experimental Chemotherapy", R.J. Schnitzer and F. Hawking, Academic Press, New York, 1963, Vol. 1, Chapter 20.
14. T.E. Pawlett, J.R. Agric.Soc.Engl., 1845, VI, 361.
15. R.J. Curtis, Chem. Ind.(London), 1987, 270.
16. H.O. Mönnig, "Veterinary Helminthology and Entomology", Bailliere, Tindall and Cox, London, 1934.
17. R. Perren, Vet.Hist., 1989/90, 6, (2), 43.
18. D.W.F. Hardie and J.D. Pratt, "A History of the Modern British Chemical Industry", Pergamon, 1966.
19. P.B. Capstick, Vet.Hist., 1981/82, 2 ,(2), 47.
20. "Animal Health and Nutrition Monitor", County NatWest Securities Ltd., 1990, (16), p.2.
21. T.W.M. Cameron, "The Parasites of Domestic Animals", Adam and Charles Black, London, 1951.
22. G.T. Brooks, "Chlorinated Insecticides", Vol. 1 and Vol. 2, CRC Press, Cleveland, Ohio, 1974 and 1975.

23. N.F. Baker, "Experimental Chemotherapy", R.J. Schnitzer and F. Hawking, Academic Press, New York, 1963, Vol. 1, p.913.
24. R.J. Hart, Reference 2, Chapter 31, p.585
25. B.H. Palmer, et al, Proceedings of The Third International Congress of Acarology, 1971, p.687.
26. M. Elliott, "Synthetic Pyrethroids", ACS Symposium Series 42, American Chem. Soc., Washington, 1977.
27. V.K. Stubbs, C. Wilshire and L.G. Webber, Aust.Vet.J., 1982, 59, 152.
28. W. Stendel and R. Fuchs, Vet. Med. Rev., 1982, (2), 115.
29. L.B. Townsend and D.S. Wise, Parasitol. Today, 1990, 6, (4), 107.
30. W.C. Campbell, Parasitol. Today, 1990, 6, (4), 130.
31. R.S. Rew and R.H. Fetterer, Reference 2, Chapter 16, p.321.
32. S. Marriner and J. Armour, Reference 2, Chapter 14, p.297.
33. D.H. Bliss and R.M. Jones, Vet. Parasitol, 1983, 12, 219.
34. R. Baker and C.J. Swain, Chem. in Britain, 1989, 692
35. G. Albers-Schonberg, et al, J.Am.Chem.Soc., 1981, 103, 4216.
36. M.Pringle, "Developments in Animal Drug Delivery", V&O Publications Ltd., 1986.
37. G.C. Cook, Parasitol. Today, 1990, 6, (4), 133.
38. P.J. Waller, "Resistance in Nematodes to Anthelmintic Drugs", CSIRO, Melbourn, 1985, p. 1.
39. J. Marchant, "Animal Pharm. Review 1990", P.J.B. Publications Ltd., 1991.

Contributions of Chemistry to Food Consumption

Harold Fore

7, THE CEAL, COMPSTALL, STOCKPORT, CHESHIRE SK6 5LQ, UK

Preface

It is said that man's need for food has shaped world history. This paper is concerned with scientific progress during the past 150 years which has resulted in better understanding of this need and in extended means for meeting it.

Selected events are identified as milestones, some deriving from many years of painstaking research, others from inspired innovation. Reference is also made to human and political factors which have influenced progress.

My sincere thanks are due to former colleagues and contacts (named under Acknowledgements) for suggestions used throughout the paper but I accept full responsibility for the final choice and emphasis.

I also wish to acknowledge the influences of the late Edgar Fore, my father and member of this Society, and of the late Professors Hilditch and Alan Morton without which I would not be in a position to prepare this paper, and to thank the recently late Alfred Williamson and the Industrial Division for the opportunity to present it.

The first quinquenium 1841-1890

Some early personalities

My invitation came from the North West of England, an area with a reputation for innovation. It was in Liverpool, but four years before The Chemical Society existed, that the British Association's annual meeting asked Justus von Liebig for a report on the state of organic analysis. Liebig put his own interpretation on the request and provided a report on "Chemistry in its applications to agriculture and physiology". In 1847 he published "Researches on the chemistry of food" which opened up the fields of food chemistry generally and meat science in particular; a major milestone.

When this Society was nearing its Ruby anniversary a group of industrial chemists in the North West sought to create a professional organisation with interests specifically in industrial chemistry and, in 1879, a Chemical Society for South Lancashire was established at Widnes. This move led directly to the inaugural meeting, held at Burlington House with the hospitality of The Chemical Society on 4th April 1881, of the Society of Chemical Industry. Later that year the Liverpool Section of the SCI was established. Forty-five years later a founder member of this section, Professor Campbell Brown, endowed a Chair of Industrial Chemistry in Liverpool which was probably the first such department in a *science* faculty of a university, and the appointee, Professor Hilditch, commenced a 25 year study of the chemistry of oils and fats, another undoubted milestone.

A further early milestone was the publication by Louis Pasteur in 1857 of "Une memoire sur la fermentation appelée lactique", widely acknowledged as the beginning of food microbiology. While both Liebig and Pasteur made invaluable contributions they also retarded further advances as, in their day, it was very difficult for lesser mortals to gain acceptance of observations which conflicted with the teachings of those held in high esteem. Liebig's misconceptions that protein was absorbed intact into the body and was the fuel of muscles took decades to dispel despite Edward Frankland's work on the bomb calorimeter in 1866, the use of which provided more precise energy balances and ultimately proved Liebig wrong. Pasteur's contention that enzymes were not "separable from life" was not effectively countered until Büchner unequivocally refuted it by producing a cell-free extract in 1897.

Liebig's 1840 report led John Lawes, a wealthy landowner, to engage Joseph Gilbert, a chemist trained by Liebig, to test the ideas on a practical scale as early as 1843 when Rothamsted Experimental Station, the first centre of agricultural research in Britain, was founded. Although these ideas were rapidly put into practice it was 40 years before the largely conservative farming community began to recognise the value of the work, by which time British agriculture was feeling the effects of heavy imports of grain. The first centre for food research in Britain, the Low Temperature Research Station was not established until 75 years after Rothamsted, and then at public expense and only as a consequence of wartime (1914-18) food importation problems.

Other early contributions

Returning to early milestones, it was in France 190 years ago that Appert invented the hermetic sealing of foods, using glass containers. Metal cans were first used by Hine and Durrant in Britain and in 1812 the food canning industry started when Bryan Donkin's factory in

Bermondsey produced foods sealed in iron containers. The Admiralty introduced canned meat initially as a component of shipboard medical supplies and, in 1845, canned foods became part of regular Royal Naval rations with the result that other canneries opened. While food microbiology was developing in Pasteur's hands the Admiralty found they had a food poisoning problem on their's. A select committee traced this problem, associated with canned foods, to the introduction of larger cans; at that time the significance of heat penetration and the importance of killing bacteria had not been recognised. Canning eventually changed the availability of foods probably more than any other food process. Canned food as we know it today owes much of its success to the breadth of chemical science needed to support the development of tinned steel, lacquers, non-ferrous cans etc.

The year that the Chemical Society was formed saw the start of large scale ice-making which enabled wider use of ice in food preservation, a procedure still used in the landing and distribution of fresh fish. Ammonia-based refrigeration was unsuitable for shipboard use and the first shipment of frozen meat from New Zealand in 1881 was made possible by the Bell and Coleman system of refrigeration based on the expansion-cooling effect using air. Bell was a Glasgow butcher and his friend Coleman an engineer; an example of interdisciplinary activity, a subject to which I return later

France was also the origin of the first of the only two wholly manufactured foods consumed today. Mége-Mouries, specifically at the request of the French navy, produced a cheap substitute for butter and, as did Appert, he received an award from a military source, in his case for margarine from the hands of Napoleon III. The second such food, Quorn, is referred to later.

Late in our first quinquenium came another analytical milestone, Kjeldahl's distillation of nitrogen as ammonia from acid-digested organic matter. Myriad are the forms in which the original concept has been and still is used, for example to determine the protein content of foods (protein = nitrogen times a factor ranging from 5.7 to 6.4, commonly 6.25, depending on the type of protein being measured).

The second quinquenium 1891-1940

Early in this period two simple reactions were identified which were to have far reaching significance.

Hydrogenation,

The simple addition of hydrogen to the double bond in a hydrocarbon chain catalysed by finely divided nickel is the basis of processes used in the edible oil industry to produce fats with the various technological properties required in food products. The important milestone here was

Normann's 1903 patent which permitted commercial application to substrates in the liquid phase with hydrogen under pressure. However, the original work was that by Sabatier and his research student Senderens, commenced some six years earlier, which demonstrated hydrogenation in the gaseous phase using freshly reduced nickel oxide. Sabatier was eventually awarded a Nobel prize in 1912.

And the Maillard reaction

That year saw the identification of another simple reaction whose ramifications influence the quality of many foods and continue to be the basis of scientific symposia. Louis-Camille Maillard, yet another Frenchman, was seeking conditions milder than those used by Emil Fischer for the condensation of amino acids into peptides. He first tried glycerol as a condensing agent but when he moved on to sugars he found that their aldehydic nature resulted in a much more intense effect on the amino acids; so was born the Maillard reaction. Maillard, like some other chemists making chance observations, was quick to link his with those reported four years earlier on the production of "flavouring and colouring matters.... When...amino compounds...are heated at 120-140°C with sugars". This earlier report by Ling of the Sir John Cass Institute was presented to the Institute of Brewing at the Criterion Restaurant, now a listed building being completely rebuilt, just down the road from Burlington House. Although Maillard soon recognised the formation of a Schiff base following the initial addition reaction, it was 36 years before the subsequent Amadori rearrangement was demonstrated on the route to the melanins which are responsible for the quality loss in some foods due to browning of their colour. This led Hodge in the USA to draw up a scheme showing how the development of brown colour in foods follows from interactions between aldehydic and amino constituents.

Maillard can have had little idea of the extent to which food research based on his reaction would expand. The production of carbonyl compounds as a result of lipid oxidation provides just one example of the implications which the Maillard reaction has for food quality and, as has been recognised in recent years, for food toxicology. The ostensibly adverse consequences of the browning reaction initially obscured beneficial effects in flavour terms and the more recently noted anti-oxidant properties of the products.

In the Maillard reaction and its consequences we have a prime example of the intellectual challenge presented to the food scientist and technologist seeking to provide safe, enjoyable food to keep us, only incidentally, in good health.

The Emergence of nutrition

The second quinquenium coincided with the development of nutrition as we know the subject today. Graham Lusk set the scene in 1906 with "The elements of the science of nutrition", such a classic that the fourth edition was published as recently as 1976. In 1906 also, Gowland Hopkins recommended chemists to seek, in plant or animal tissues, unknown substances with unknown properties; he was after what he called "accessory factors of the diet". He based his theory for the existence of the vitamins on a number of earlier observations, for example that made five years previously in the Dutch East Indies by Eijkman and Grijns leading to 'water-soluble B', distinguishable from the factor identified by Hopkins himself in 1912, and called 'fat-soluble A' by McCollum and Davis in the USA. Six years later Mellanby identified another fat soluble factor, designated 'D' and major advances over the next 40 years led to the isolation, identification and synthesis of virtually all the vitamins recognised today. This international thrust in organic/bio-chemistry seems to have progressed virtually uninterrupted between 1914 and 1918. By contrast, although Drummond's group had demonstrated the relationship between vitamin A and carotene in 1920, it was a further 37 years before Moore at the Dunn Nutrition Unit was finally able to prove that carotene was in fact the precursor of vitamin A, by then called retinol.

Vitamin research clearly illustrates how advances in the first half of this century became increasingly dependent on teamwork and contributions from many quarters, including industry, and indeed upon rivalry; vitamin B_{12} was crystallised independently within three weeks by teams in Glaxo UK and Merck USA. Milestones are now rare, hectometer pegs are commonplace; one might even add that chemists and road engineers both now work with machines and computers rather than hands.

The year Gowland Hopkins discovered vitamin A saw the death of a wealthy merchant, Sir William Dunn, and it fell to Sir Jeremiah Colman of the mustard firm, as chairman of the will trustees, to devote a substantial sum to charity. Through Sir Jeremiah's connections, including Sir William Hardy (by now secretary of The Royal Society) and the secretary of the Medical Research Council, the Dunn Nutritional Laboratory was set up in 1927 in Cambridge, with Dr L J Harris, a Liverpool man and Manchester chemist, as director, to work on the fundamental biochemistry behind nutrition. I return later to nutrition and the Dunn but it is worth mentioning here that Dr Harris was instrumental in the formation of the Nutrition Society which also celebrates a notable anniversary this year, its 50th.

Teamwork

The problems of the nation's food supply during the 1914-18 war led the Department of Scientific and Industrial Research (DSIR) to set up the Food Investigation Organisation (FIO) whose principles in the selection and allocation of priorities to research were based on the:

- close co-operation with the user industry from conception to fruition
- need for a proper balance between applied and long term research
- prevention of the spreading too thinly of available effort.

Mr W B Hardy directed the work of the FIO and was among the first scientific administrators to establish clearly the value of gathering together on one site people of different disciplines and organising them as a team to work on specific problems. This began at Cambridge where the Low Temperature Research Station (LTRS) was established in 1922 with unique facilities for research at low temperatures. Several years later the Ditton Laboratory in Kent was set up to work on fresh horticultural produce and the Torry Research Station in Aberdeen to work on fish. The DSIR also established three independent food research associations at Leatherhead, Chipping Campden and St Albans (later moved to Chorleywood) serving commodity sectors of the food industry.

An early success and milestone from the teamwork engendered by Hardy at LTRS was the wide adoption of Kidd and West's system for the controlled atmosphere (CA) storage of fruits. This was achieved without the full understanding, won much later, of the significance of ethylene in fruit ripening, a consequence of the work of Gane providing, one presumes, the 'proper balance between applied and long term research'. Tributes are due to the philosophy at the Station, and to Kidd in particular, because CA storage found immediate application in the horticultural industry. Many of you will, I am sure, have at some stage known two individuals like Franklin Kidd and Cyril West, so different in their personalities yet working so productively together; remarkably they rarely socialised outside the workplace. Kidd, the extrovert and leader, succeeded Sir William Hardy as head of LTRS and was also director of the FIO until the transfer of the LTRS to the Agricultural Research Council in 1958. The ARC later replaced the Station at Cambridge with the Food and Meat Research Institutes at Norwich and Bristol respectively. The Bristol laboratory was recently closed as a result of cuts in public research budgets but the high quality of the work continues, as demonstrated by the award to two scientists in the surviving centre at Norwich of both the Senior and the Junior Medals by the Food Chemistry Group of our Society.

FIO workers made many contributions to the elucidation of factors affecting stored foods including the ramifications of the Maillard reaction and the complexities of fat oxidation. Another such factor, is starch crystallisation (retrogradation) which results in a marked change in crumb texture in baked goods resulting in staling of bread and similar goods. Our understanding of this type of spoilage owes much to the basic carbohydrate science developed by Haworth and Hirst also in the 1920s, building on the much earlier and brilliant work of Emil Fischer, converting his straight chain structure for the sugars into the cyclic version familiar today. This major advance was eventually extended with contributions by Whelan and others to the structure and enzymic modification of starch and glycogen, and by Shallenberger to the conformational features that contribute to sweetness. Sir Walter Haworth (Chemical Society 1944-46) and Sir Edmund Hirst (1956-58) are two of the three chemists mentioned in this paper who held presidencies in this Society; the other was Sir Edward Frankland (Royal Institute of

Chemistry 1877-80) whose bomb calorimeter was referred to earlier; he was better known for contributions in the context of public health to water analysis, a major food constituent.

Despite what I have said about the achievements of FIO workers one should note that, although they probably understood the rapid freezing process applicable to vegetables, it was the Americans led by Clarence Birdseye who developed it, and the necessary downstream technology, to give us frozen foods.

The impetus of war

In 1939 the Medical Research Council published a report on the composition of a wide range of foods. This report, intended to assist in the treatment of diabetes, was the origin of today's more widely applicable food composition tables known as 'McCance and Widdowson's The Composition of Foods' now published for the Ministry of Agriculture Fisheries and Food by this Society. That same year , forseeing problems of contamination of wartime food supplies with poison gas, the government appointed Professor Jack Drummond adviser to the Minister of Food. Whether syrendipic or not this choice was one factor, alongside rationing and price control, in ultimate victory. Like Liebig, Drummond, also a chemist, had his own ideas about the area in which his advice could be used, indeed he had anticipated the need and immediately presented a memorandum on the value of co-ordinating all scientific investigations into food problems, followed quickly by a second "on certain nutritional aspects of the food position". The successful feeding of this nation during the war, and sectors of others in the aftermath, was significantly dependent on his input and ability to stimulate others into action.

The scientific and technical base on which Drummond was able to draw derived in part from the LTRS and the other food research bodies established by the DSIR referred to earlier Success in feeding the nation at war stemmed from Sir Jack Drummond's (knighted in 1944) unique breadth of contacts in these and many other places. Early on he had to visit the USA and obtain information to determine the intakes of vitamins and minerals to be used as a basis for maintaining a healthy diet; at that time no such criteria had been developed in Britain. Nonetheless, much chemical analysis and other work was necessary under the auspices of the Ministry of Health to provide a new table of food composition as a basis for directing the optimal home production and minimal essential importation of foods to meet the needs of all sectors of the population and those of the armed forces based in Britain.

The third quinquenium 1941-1990

Food analysis and legislation

Because the legitimate sale of food is so dependent upon its precise composition the importance of food analysis cannot be overemphasised. Not only has it been essential in the fundamental advances referred to already, it is also crucial in the development of manufactured foods and the quality control of all foods. Less obvious is the dependence upon food analysis of the food processing industry , as we know it today, wherein the chief chemist has metamorphosed into that Jack of all trades, the food technologist. Indeed, had the chemical data been complemented by comparable physical data on food substances the food process technologist would be as well equipped as the chemical engineer in designing processes. The need for data on what the food engineer calls 'the working substance', our food if you please, led to an EEC-funded programme to generate it. However, in the event, most of the effort had to be devoted to method rationalisation - a *sine qua non* of the chemist. The transition of food manufacture from cookery on a scale at which food quality may suffer, to a process industry, with the benefits such an approach has to offer, has been inhibited in part by this lack of physical data. An example of the need for such data is in the heat transfer modelling required to arrive at a composite dish or meal which will reheat satisfactorily in a microwave oven for the convenience of the busy consumer.

In transferring and converting the output of agriculture, horticulture, fisheries, and now biotechnology, to consumer products, the food processing and retailing industry's ability to monitor raw materials, processes and products is paramount. Our third quinquenium has seen dramatic changes in the requirements on the human food chain, with the production, the preservation and conversion, and the sale of food merging in reponse to societal changes which have become particularly rapid in recent decades. Modern techniques such as gas chromatography, mass and near infrared reflectance spectrometry, to mention only those better established, and their ability to detect and often measure the hitherto undetected, have solved many monitoring problems. But it has to be admitted that almost as many new problems have arisen from the sensitivity of these techniques.

In the 1940s Hilditch's group spent days separating triglycerides by fractional crystallisation from acetone in flasks at -50°C but gas chromatography soon thereafter achieved this in seconds with milligrams. Nearby, Alan Morton's group, applying spectrography to biological problems, had to persuade the Biochemical Journal to allow the use of "micromilligram", whereas now the nanogram is large beer alongside pico- and femtogram levels causing concern over residues in food. Today's 'not detected' is tomorrow's legislative upper limit, the general public become understandably concerned, food activists have a field day, a politician misinterprets a brief and the media have difficulty, even if the will, in giving balanced reports. In retrospect, the problem of aflatoxin in peanuts, which started life as a disease in turkey poults, was dealt with remarkably quickly without much scaremongering although, in the longer term, it has resulted in the monitoring of foods for mycotoxins.

Such developments have led to the extensive and, in some situations, very complex legislative framework within which the food industry operates today. Again one cannot overemphasise the intellectual input and resourcing required of both industry and government to operate within this framework, which now has to be compatible throughout the EEC. We have come a long way since the 1860 Food & Drugs Act whose sole food purpose was to counter adulteration; current legislation is far more comprehensive and the 1990 Food Safety Act seeks widespread control of food hygiene. The 1860 Act failed because vested interests inhibited the appointment and activities of analysts; the Achilles heel of the 1990 Act is the limited resources available to local authorities adequately to monitor the multifarious food purveying activities caught by the legislation.

Today, the consumer expects, or takes for granted, foods which are:

- produced without offending the purchaser's sensitivities
- fresh from many parts of the world
- free from allegedly harmful substances
- in minimal need of preparation for cooking or eating
- compatible with the cook's flair, or lack thereof
- labelled to provide compositional and nutritional information.
- presented attractively and hygienically but not so as to mislead.

Legislators need not only to determine, for example, whether and how irradiated foods may be sold and described, but also to define such words as 'natural' and 'fresh'. Many of us are pleased to purchase 'fresh' milk but would be very displeased if it were not pasteurised. In fact pasteurisation was first used in 1890 as a means of increasing the shelf life of milk; six years elapsed before its benefits in terms of hygiene were recognised. Pasteurised milk was widely introduced in this country just before the 1914-18 war. The safety of food irradiation, the early process development of which was done at the LTRS, has been more thoroughly investigated than pasteurisation but there is much opposition to its introduction, partly and ironically because of the idea that its main benefit is in increasing shelf life and, allegedly therefore, to the food industry. An obstacle to satisfactory legislation has been the need for a cheap routine test for identification of irradiated food; sophisticated techniques such as high field, high resolution NMR have proved necessary to detect the very low levels of radiolytic products it contains.

Government, recognising the need in determining effective legislation for input by industrial and academic chemists and other specialists, established an advisory committee system in 1923. This has been adapted progressively to changing needs, with the result that since

1983 we have had the Food Advisory Committee, in which the consumer interest is now directly represented, advising the Ministers responsible for food and health on many aspects of food legislation, including those referred to earlier. These committees have published about 125 reports since 1923 but, with the need for advice in the context of EEC negotiations and other factors, more *ad hoc* advice and fewer formal reports are now generated. The development of new technology in the food area has led to an Advisory Committee on Novel Foods and Processes.

A recent development in food legislation is the stipulation of the form in which nutritional information should be given for food products where the vendor volunteers to provide such data. It is interesting to note what Graham Lusk said nearly 80 years ago about the provision to the consumer of nutritional information on foods: "Who will give him this...? Will the manufacturer? No, not unless he is forced...by law...to label his can 'This contains x calories, of which y% are protein of grade C'".

Biopolymers

While the earlier analytical techniques contributed immensely to chemical advances in carbohydrates and vitamins, advances in lipids and proteins awaited the several developments in chromatographic techniques already mentioned and the works of Chibnall, Sanger, Kendrew and Perutz. More recent techniques, such as the several variants of NMR, have given us much greater understanding of molecular fine structure and behaviour of complex molecules and also of that most abundant constituent of foods, namely water. A recent paper has shown the value of using a combination of such techniques, in this instance for the study of amylopectin crystallisation during bread staling, or the retrogradation of starch as it was called earlier in this paper.

In parallel with these advances the term biopolymer has been introduced for complex structures composed of one or more proteins, lipids and/or carbohydrates, in or with which water often plays an essential role in technological performance as well as in living tissues. Thus the significance of water has advanced beyond that as a mere constituent, or indeed that as 'water activity' in food preservation. Water is now seen to have an all important role in molecular conformation and behaviour in the complex mixtures of biopolymers which typify much food consumed today. Without the foundations provided by the teams led by Howarth, Hilditch, Chibnall and others on basic structure and more recently on fine structure, such as that on meat collagen and elastin by Partridge and Bailey, these recent advances would not have been possible.

Fundamental study of food emulsions and foams is another area from which food technology is gaining significant benefit and, with better understanding obtained from these and other studies, less empirical approaches to food preservation, processing and the use of additives will be facilitated.

Nutrition education

In 1945 Lord Woolton, another Liverpool man and Manchester graduate, expressed concern at the non-existence of a professor of nutrition in Britain in contrast to the USA. A move to establish one in the University of London initially failed and Dr Harris travelled from the Dunn Nutritional Laboratory to Kensington 2-3 times a week to lecture on nutrition in the Department of Physiology at Queen Elizabeth College. When John Yudkin became Professor of Physiology there in 1946, he had to overcome objections in principle to a syllabus for a degree in nutrition in a *science* faculty. These related to the breadth and vocational nature of the syllabus, the unsuitability at degree level of studies on the effect of cooking on food, not to mention allegations of trespass by biochemists and physiologists. Seven years of determined effort resulted in the first nutrition degree course in Europe and, the following year, Yudkin transferred to a chair in nutrition alongside that in physiology. The syllabus had been improved in at least one respect, it encompassed "preparation of food for human consumption" as distinct from "cooking". Ironically there is some truth in the belief that the very success of Sir Jack's wartime efforts led to the feeling that few nutritional problems remained and there was no future for an academic resource. The consequences of the increasing average age and affluence of our society has since put paid to that idea. The relative recency of the change in emphasis at the, now, Dunn Nutrition Unit to investigation of specific nutritional problems further reflects the obstacles faced over the years by those keen to develop nutritional science, not helped by the tragic murder of Sir Jack and his family in 1952 and the untimely death of Sir Alistair Frazer in 1969, two years after he was appointed Director General of the new British Nutrition Foundation. The future of the Dunn Nutrition Unit is now dependent in part on funding by the food industry.

Food processing and manufacture

University teaching and research in food science and technology faced problems analogous to those described for nutrition but there are now five departments and several polytechnics offering degree courses. Furthermore, concurrent with this Congress is a conference in Brussels on the education of food scientists, engineers and technologists. Nonetheless, the scientific and technical input for the harvesting, preservation, processing, manufacture, storage, distribution and retailing of food is most effectively provided by food technologists working alongside people from a wide range of disciplines. The need for this variety of input is well illustrated by the resumé on the next page of milestones contributing to food consumption over the 150 years of our Society. Seven of the last nine have yet to be expanded.

Milestones 1841-1990

[1800 Appert - hermetically sealed foods in jars]
[1812 Donkin - hermetically sealed foods in cans]
1841 commercial ice making
1847 Liebig - father of food analysis
1857 Pasteur - father of food microbiology
1869 Mège-Mouries - margarine patent
1881 Bell and Coleman - expansion-cooling shipboard refrigeration
1883 Kjeldahl - nitrogen determination

1897 Büchner - cell-free extract with enzymic activity
Sabatier and Senderens - double bond hydrogenation
1902 Fischer - projection formula for sugars
1903 Normann - hydrogenation catalyst patent
1906 Graham Lusk - 'The elements of the science of nutrition'
1911 Tswett - liquid/solid chromatography
1912 Maillard - non-enzymic browning reaction
Gowland Hopkins - 'accessory factors of the diet'
1923 Deptl Ctte on Preservatives and Colouring Matters in Foods
1926 Hilditch appointed Professor of Industrial Chemistry
Howarth - ring formula for sugars
1926 Kidd and West - controlled atmosphere storage of fruit
1928 Clarence Birdseye - rapid freezing
1939 McCance and Widdowson - The Composition of Foods
Professor Drummond appointed adviser to Minister of Food

1946 Chibnall - protein molecular composition
1948 Moore and Stein - liquid/charged gel electrophoresis
Proctor and Gamble - transesterification patents
1952 James and Martin - gas/liquid chromatography
1960s Josel - supercritical fluid extraction
1961 Chorleywood bread process
1970s properties of water
foods by extrusion cooking
1973 mycoprotein on a pilot scale
1980s immunochemical analysis of food
sensory analysis

Edible oils and fats consist of mixtures of triglycerides in which the three fatty acid moieties have been directed on to the glycerol molecule systematically by one or other enzyme system. The technological properties of these natural products, and specifically the melting point range of the triglycerides, can be changed by controlled randomisation of the acyl groups. Although this transesterification, as it is often called, was first achieved in the 1920s, high temperatures were required. It was 25 years before the commercial potential was realised with the introduction of catalysts effective at temperatures where manipulation of the reaction through product removal or other means became possible and many patents issued between 1946 & 1948. Application of the process is exemplified in the production of fats with better shortening properties for baking purposes than those of the natural product. Transesterification can provide fats which solidify with a very fine crystal network capable of immobilising oil and droplets of aqueous phase. This network contributes, in the case of batters, to the creaming behaviour familiar to cooks.

Today's wrapped sliced loaf, which often takes a lot of stick yet meets many requirements very well, is produced using a process developed by the Flour Milling and Baking Research Association at Chorleywood. Satisfactory bread depends in part on adequate 'development' of the flour gluten and, in the traditional process, this stretching of the gluten molecules is brought about by the expanding volume of yeast-produced gas. In the Chorleywood process the gluten is developed by mechanically beating into the dough a precise amount of energy with the dough at an initial temperature which leads to the required final temperature following the increase during energy input. It is necessary to incorporate a more active yeast preparation to produce, during the shorter process time, the desired porous structure and flavour in the bread. This is an example of where technological push rather than market pull led to a new product; but only because Lord Rank was persuaded to take the risk and develop the market.

It was also Lord Rank's faith in scientific endeavour which enabled a start to be made on the second wholly manufactured food presently consumed, nearly 100 years after the first, margarine, was patented. The idea was to exploit the fibrous properties of microfungi to provide inherent textural properties in a fermentation product, a property lacking in all such products developed for food use hitherto. The project started in 1963 with the isolation of a suitable species of Fusarium which, by 1973, had been successfully grown on starch liquor, a by-product of gluten manufacture, under careful control in 1000 litre fermenters. The wet mycelial mass had then to be formulated into a basic food suitable for human consumption, now called "Quorn". It has been available since December 1985 in a wide variety of products sold mainly as own label items by larger supermarkets. A product of conventional biotechnology, but dependent on much microbiological, biochemical, technological and toxicological input, Quorn was cleared, for marketing by Marlow Foods, by the government in October 1985 on the advice of the Advisory Committee on Novel Foods and Processes. This Committee considers that "any substance promoted as a replacement or an alternative to a natural food should be the nutritional equivalent in all but unimportant aspects of the natural food which it would simulate". This exacting requirement Quorn has clearly met. The next quinquenium will determine whether my choice of mycoprotein as a milestone is justified.

Another technological process which has been used to convert cereals into edible products, extrusion cooking, was originally developed on a purely empirical basis but in recent years it has been the subject of systematic study. The technology derives from the plastics industry and involves precise process control of extrusion, at very high pressures and relatively high temperatures, of a mixture formulated such that the extruded material is the final consumable product. The so called snack product market owes much to this process.

Concern about residues in foods, perhaps in flavourings in particular, has created a need for a solvent whose residues, if any, can be discounted in food safety terms. A gas, such as carbon dioxide, above its critical temperature and pressure, becomes a fluid with solvent characteristics dependent upon the temperature/pressure combination. Extraction of a substance from a substrate and its recovery from solution can be achieved by appropriate differential combinations of temperature and pressure. Decaffeinated coffee can be produced using supercritical CO_2 in place of methylene chloride. Present process costs limit its application but the area of flavourings is another where supercritical extraction should prove economically viable.

The introduction of immunochemical methodology has provided a further important step forward in food analysis.

And the next 50 years?

Elizabeth Bird, 150 years ago, reacted adversely to the albumin in eggs. Alfred, her husband, invented egg-free custard powder and for many years Bird's Custard was coloured with tartrazine. It is interesting to note that at that time tartrazine was the only food colouring which appeared on lists of permitted food colourings throughout the world, where such lists existed. In recent years, because some people react adversely to tartrazine, it has been replaced in some foods and in Bird's Custard annatto is now used instead. Despite the effort devoted to ensuring safety of food a low level of adverse reaction to food among consumers generally is recognised. Some is due to substances naturally present in food; Mrs Bird was reacting to egg albumin. But some is attributable to substances given official clearance for food use. Adverse reaction to diet is a complex area, hard evidence is lacking and there is much scepticism. The complexity is exemplified by the headaches, known to be associated with excessive coffee consumption, which have also been shown to occur following a switch, at moderate intakes, from ordinary coffee to decaffeinated coffee. I believe that ensuing decades will see more serious attention to such problems and to more comprehensive cost/benefit analysis of the use of food additives.

Endpiece

Our 150 years has seen the western world move from a situation in which the majority had cause to consider when and from where they might get the next meal regardless of quality, to one where the majority have a wide choice of almost impeccable food products of which they tend to eat too much of an ill-advised selection and question the safety of some. In the UK, expenditure on food as a proportion of the household total has fallen markedly; in 1938 it had reached about 27%, in 1977 17% and recently the Minister quoted <12%. This is a reflection of a British attitude deriving from relatively cheap food from the Empire and John (later Lord) Boyd Orr's pronouncements in the late 20s on behalf of the *poor*. Boyd Orr was the director of the Rowett Research Institute. The present director, Dr Philip James, also has expressed concern, but for very different reasons, about contemporary diet and has played a significant role in moves to implement dietary guidelines to combat diseases of our *affluent* society This further illustrates the point evident elsewhere in this paper, plus ça change, plus c'est la même chose.

Chemical analysis of food was our first milestone but at our Sesquicentenial Congress it has to be said that, while taste evaluation has been used for many years, only recently have a variety of statistical techniques refined it to such an extent that the term "sensory analysis" is justified. Such analysis should, after all, be the most appropriate tool in food quality control. The most significant of what Gowland Hopkins called the 'accessory factors of the diet' must be the appeal of the food to the consumer; that which is not eaten has no nutritional value. I have therefore selected sensory analysis as a milestone to complete our review period and end this paper.

This review had to be selective and could not embrace a number of areas none the less important to food consumption. These include: agricultural production, wherein chemistry has contributed via fertilisers, agrochemicals and veterinary products and has facilitated gene manipulation in precision breeding for desired characteristics in plants such as high and low glutenin levels in wheat; biotechnology, including bioengineering and the provision of a supply of potable water for food processing as well as for general consumption; the development of semiconductors without which computer control of food processes would not be possible; the development of plastic packaging without which the hygienic retailing of foods on open display would not have arrived; chilled distribution of foods.

Acknowledgements

Contributions of various kinds were made to this paper by: Miss D F Hollingsworth OBE; Professors A E Bender, J F Diehl, G A H Elton CBE, J Hawthorn, A W Holmes, T A Kletz, R A Lawrie, W R Morrison, H E Nursten, P Richmond, D S Robinson, M P Tombs, A G Ward OBE and J Yudkin; Drs A W E Axford, G G Birch, T Sharp, R O Sharples, R G Whitehead and J J Wren; Messrs M R Alderson, J R Blanchfield, R Headland, J Horton, P Sulsh and A Turner OBE; Dr G R Fenwick and the committee of the Society's Food Chemistry Group; and the information sections of some centres of food research. All were much appreciated.

Clothing

Chemistry and the Development of Natural and Synthetic Fibres

David A. Hounshell

DEPARTMENT OF HISTORY, CARNEGIE MELLON UNIVERSITY, PITTSBURGH, PENNSYLVANIA 15213-3890, USA

1 INTRODUCTION

The textile arts are among the oldest pursued by humankind. By the eighteenth century, they had been brought to a high degree of perfection, as the finest textiles from that period preserved in museums in Italy, France, England, and other European nations make abundantly clear. Artisans had learned to spin, weave, manipulate, dye, and finish natural fibres of all types; their knowledge of these fibres and processes was largely experiential and tacit. No highly developed, formally articulated body of knowledge governed their craft. Rules of thumb and special recipes characterized the artisans' work.

Yet the mechanization of textile production--associated so intimately with the "Industrial Revolution" that historians have recently come to question and the "Consumer Revolution" that is now commanding so much of their attention--created an entirely new dynamic in the textile arts. Pressure to know and manipulate fibres better came both from production side needs (e.g., greater throughput) and from the demand side (e.g. the desires of consumers for "whiter whites," greater uniformity and predictability in textiles, new and faster colors, and new finishes). These changes at the end of the eighteenth century proved propitious, for the "Chemical Revolution," about which historians of science still speak and write, was just unfolding. Chemical knowledge was soon joined with textile practice, initially only in a few important instances but then, beginning about the time the Chemical Society was established, with increasing frequency. During the twentieth century, chemistry came to play a vital role in the textile industry. Without chemistry, there would be no man-made fibres industry. The textile industry, the natural and synthetic fibres industries, and related institutions have during much of this century supported an extensive amount of chemical research, which not only contributed to new and improved products and processes, but in some important instances added substantially to the body of chemical knowledge.

This paper explores the changing historical relationship between chemistry and natural and synthetic fibres roughly during the 150 years spanned by the Chemical Society's history. The challenge in writing this paper has been two-fold. First--and most obvious--is to contain the paper within the proscribed length. Second--and more critical--is to ensure that the paper emerges as something other than a recitation of "Progress." Therefore, the milestones treated in this paper are intended to illuminate the process of historical change and the dynamic interaction of humans, their institutions, and their science and technology rather than to mark steps on an endless road of progress.

2 MERCERIZING

When the Chemical Society was founded in 1841 amidst the national debate over chemical science versus chemical practice, chemistry already played a significant, though by no means a central or a sustained, role in the British textile industry, which in terms of its size and diversity was unequalled in the world. Without the development of chlorine bleaching in the late eighteenth century--a wonderful, international story of the rapid application of new chemical theory--the British cotton industry probably could not have achieved the level of output and quality that it gained by the 1840s. One of the many British industrial chemists who worked feverishly on chlorine bleaching, Charles Tennant, is best known for his development of chlorine bleaching powder. But Tennant's partner, Charles Macintosh, who contributed substantially to that development, went on to make a household name for himself with the development of the rubberized, waterproof raincoat that bore his name--the Macintosh, surely one of the first branded products. Macintosh's process of coating fabric with rubber grew out of his efforts to obtain ammonia from coal tar.

Neither chlorine bleaching nor the Macintosh, however, possessed the implications implicit in the discovery and development of the first milestone in our history of chemistry and fibres: cotton mercerizing. The word "mercerize" derives from the discoverer of this process, John Mercer, a Lancashire calico printer who had initially pursued the craft of his father, a handloom weaver. By 1844, however, when he observed that caustic soda swelled cotton fibres and shrank their length, Mercer had been in the dye business since 1809. Sometime earlier he had become intensely interested in the more theoretical aspects of his trade. In 1841, Mercer, a founding member of the Chemical Society, met Lyon Playfair, a Scottish chemist who had recently finished his education under the famous German chemist at Giessen, Justus Liebig. Playfair, who in 1848 would draft the Chemical Society's charter, was then working as a chemist at another calico printing firm. He would soon, however, move on to what might be described as a major phase of empire building, predicated on his belief that, as Bud and Roberts put it, "applied science itself could be an academic subject."

Mercer and Playfair were part of a circle of industrial chemists who gathered weekly in a tavern at Whalley to discuss chemistry. Here Mercer chided his friend Playfair to publish substantial pieces on chemistry, not just small, occasional notes. Here Mercer first propounded his theory of catalytic action, which he presented more formally at the Manchester meeting of the British Association for the Advancement of Science in 1842. Here also Mercer pondered with his chemical friends fundamental questions about solutions of differing concentrations possessing different viscosities and mobilities, questions stimulated by at least two papers published by Thomas Graham in the new Memoirs of the Chemical Society. As Mercer's biographer recounts, Mercer suggested to his circle of chemists that one could perform an experiment whereby solutions of different concentrations could be passed through capillaries, and he predicted that mobility rates would be consistent with "different degrees of chemical hydration."

Mercer's interest in the fundamental phenomena associated with different concentrations of solutions continued to develop. He determined to try to effect "a partial separation of different hydrates by slow fractional filtration." As Mercer later stated: "For this purpose I made a filter composed of six folds of strong, fine cotton cambric, bleached, passed three times through the calender to make it compact, and poured upon it solution of caustic soda of 60 Tw[addell]. The filtration was very slow; the liquor which passed through was 53 Tw. . . . But I found my filtering-cloth had undergone an extraordinary change; it had become semi-transparent, contracted both in length and breadth, and thickened or 'fulled,' as I then termed it." Mercer fiddled around with several swatches of his "fulled" cloth but because of other pressing business dropped his inquiry until after 1848.

When he returned to the matter, Mercer demonstrated his power as a chemical thinker by asking two important questions. First, he pondered whether the "fulled" or "sodaized" cotton had undergone any chemical change or merely mechanical alteration. Second, he explored whether this change had altered the material's receptivity to dyeing and printing. Using both the microscope and chemical reasoning, he concluded that indeed the cotton had undergone a chemical transformation. Further investigation demonstrated that the strength of the cotton had actually been improved in the process and that the "mercerized" cotton's affinity for acid, neutral, or basic dyes had been substantially increased.

Mercer subsequently carried out additional, extensive analysis and conducted large-scale trials of mercerizing cotton. In 1851 he received a British patent for his process in which he essentially made eight claims for his alkali treatment of natural fibres. At the Crystal Palace Exhibition, Mercer exhibited the products of his process. As Bruce Hartsuch, author of a standard text on textile chemistry wrote, "This was the beginning of one of the most important advances that ha[s] ever taken place in the textile field. . . . An enormous amount of research has been done since that time on the action of alkalis on cotton, but not much has been added to the original list of observations made and patented by John Mercer." Mercer's work--including much of his more theoretical chemistry not discussed here--earned for him election to the Royal Society in 1852.

Although not a commercial success in the Lancashire cotton industry of the 1850s, Mercer's discovery and patent, I believe, were absolutely fundamental to subsequent developments in textile chemistry. Mercerizing, especially when it was brought into widespread use through the Lowe patent of 1890, which produced yarns and cloth of high luster, demonstrated in a dramatic way that natural fibres of all types could be chemically modified and indeed improved. Since the time of Mercer, as we shall see, the name of the game in textile chemistry of natural fibres has been chemical modification and treatment to enhance or introduce desirable properties and to reduce or eliminate undesirable properties. The very idea of chemical modification provided the basis for the rise of what initially became known as the artificial fibres business, or what is today called the man-made cellulosic fibres. More specifically, the treatment of cotton with caustic soda--John Mercer's basic process--was the first step in the process that produced what E. J. Bevan and Charles F.

Cross called "viscose," which since 1924 has been called rayon. Indeed, Bevan and Cross were working explicitly with mercerizing in 1891 when they discovered the viscose reaction.

3 RAYON

Although Robert Hooke had suggested in his Micrographia (1664) that one might be able to "make an artificial glutinous composition much resembling" the product of the silkworm and although the French scientist Reaumur wrote in 1734 (Memoirs pour servir a l'histoire des insectes) that humans, not just silkworms, should be able to spin gums or resins such as those of the mulberry into fibres, not until the middle of the nineteenth century was any serious effort devoted to developing artificial fibres. Unfortunately, the history of these efforts has never been adequately developed. What I find interesting is the contexts in which the pioneers of artificial fibres worked. Let us briefly examine them.

At the 1842 meeting of the British Association, Louis Schwabe, a Manchester silk manufacturer, presented a paper in which he described a machine that would extrude a resinous solution through small orifices to form artificial fibres. Fifteen years later, another Mancunian textile chemist, E. J. Hughes, obtained a British patent for silk-like fibres made from a mass consisting of starch, glue, resins, tannins, etc. Hughes must have been stimulated by the issue in 1855 of the first British patent on artificial fibres, that of the Swiss chemist Georges Audemars, who described how silk-like filaments could be formed below a needle pulled out of a solution of cellulose nitrate, alcohol, and ether and then wound on a bobbin.

The key to Audemars's patent was its use of cellulose nitrate, which had first been obtained in the 1830s and which led in 1846 to the early production of guncotton by Christian F. Schönbein. Widespread work on cellulose nitrate also led to the development in the late 1860s and early 1870s of the celluloid industry--the first plastics industry.

The increasing knowledge of cellulose nitrate chemistry played an important role in Joseph Swan's work in filaments for incandescent electric lights. In 1883 Swan obtained an important British patent covering the production of artificial fibres by extruding a solution of cellulose nitrate and glacial acetic acid through a tiny orifice, or spinneret, into a precipitating bath. Swan then carbonized these filaments and used them in his electric lamps. In 1884, he exhibited what he called "artificial silk" fibres at a meeting of the Society of Chemical Industry in London and a year later showed fabrics crocheted from these threads--what he called "silk made by art"--at the Exhibition of Inventions in London. Whereas numerous experimenters had by this time extruded cellulose nitrate into fibres, which were very dangerous because they were essentially guncotton in fibre form, Swan had figured out how to denitrate them, converting the cellulose nitrate into cellulose hydrate. Because Swan used his process only to make electric lamp filaments, his work is often overshadowed by that of "The Father of the Artificial Silk Industry," Louis Marie Hilaire Bernigaud, later known as Comte de Chardonnet.

Chardonnet came to his work on artificial silk through the study of real silk. Born in 1839 in Besançon, France, Chardonnet studied chemistry there under the brother-in-law of Louis Pasteur. (Pasteur himself had been a graduate of the Royal College of Besançon.) In 1859 Chardonnet entered the Ecole Polytechnique and was still in Paris when Pasteur undertook his famous studies on diseases of silkworms for the French silk industry. Chardonnet in some way aided Pasteur with this study, carrying out "field" research for Pasteur in the silkworm-growing areas in southern France. Careful observation of the silkworm, its "feedstocks" of mulberry, and its spinning process, led Chardonnet, who was fully aware of the English and German work on cellulose nitrate, to try his own hand at making artificial silk. Eventually his research would be supported by the French Academy of Sciences.

In what might now seem a naive idea, Chardonnet began by using a liquor made from the twigs and leaves of the mulberry plant. He then nitrated the liquor, dissolved it in alcohol and ether, and then extruded the cellulose nitrate through a tiny orifice similar to the spinneret of the silkworm. What he obtained, of course, was guncotton in fibre form. Thus when Chardonnet exhibited fashions created from his silk-like filaments at the fabulous 1889 Paris International Exhibition, one is not necessarily speaking figuratively about how explosively they hit the world of Paris haute couture! For a while, Chardonnet silk was banned in France because of its safety problems, and the English cautioned the public about its dangers. Nevertheless Chardonnet's fibres were indeed sensational, and he quickly attracted sufficient capital to undertake their commercial production. Joseph Swan's work showed him how to denitrate his fibres, so eventually they were far safer (though still flammable). By 1891, his factory in Besançon was daily producing about 125 pounds of the very bright and shiny, though somewhat wiry fibres. By the early years of this century Chardonnet was also operating plants in Switzerland, Belgium, and Hungary.

Chardonnet obtained something of a proprietary position with his fibres based on his fundamental 1884 French patent on the process and some later important German process patents. During the 1880s and 1890s, chemists throughout Europe--and even some in the United States--were devoting increasing amounts of time, energy, and money either to the improvement of Chardonnet silk or to finding ways around the patent. Chardonnet silk possessed numerous problems, including its very poor wet strength and extreme unevenness in dyeing, owing to poor control in the denitrating process. Better process control and purer materials improved dyeing uniformity. The ca. 1890 development of the cuprammonium process by the French chemist L. H. Despaissis and subsequent improvements by several German and Swiss chemists also greatly improved dye uniformity while also yielding a fibre with better hand, or feel. By 1899, German capitalists had made possible the commercialization of the cuprammonium process through the creation of the Vereinigten Glanzstoff Fabriken A. G., which soon had plants in Germany, France, England, Wales, the United States, and elsewhere. My best estimate suggests that by 1900, more than 2 million pounds of nitrocellulose silk had been manufactured with the Chardonnet and cuprammonium processes.

Research and development activities in nitrocellulose fibres provided an important context for the work of Charles F. Cross and E. J. Bevan that led eventually to the viscose rayon process. But the textile industry's renewed work on mercerizing natural fibres provided the more immediate context for their research. Horace Lowe's patent of 1890 and its very rapid adoption within the textile industry brought about an enormous interest in cellulose chemistry and the chemical modification of natural fibres. Bevan and Cross had been students at Owens College, Manchester, in 1878-79 and had developed there a keen research interest in cellulose. After leaving Manchester, the two worked as consulting chemists while carrying out systematic studies of reactions of cellulose. The issue of the Lowe patent prompted them to study mercerizing in 1891. Within a year, they had taken mercerizing one step further by reacting cellulose more completely and with higher concentrations of caustic soda ("complete mercerizing") and then treating it with carbon disulfide, which yielded a thick, honey-colored syrup they called viscose. Although much worked remained to make viscose into the fibre product that it eventually became, Bevan and Cross's work was fundamental.

What I find so interesting about the development of viscose is the way in which contemporaries viewed the innovation and the language they used to describe processes. Very strong lines of continuity are suggested rather than the radical change that some historians of rayon have promoted. For example, in a 1910 textbook on textiles edited by the well-known British textile chemist Andrew. F. Barker, W. M. Gardner contributed a chapter entitled "The Mercerized and Artificial Fibres. . .," which suggests, to me at least, that Gardner viewed these developments as a continuum. He stressed that the "production of Mercerized Cotton [by the Lowe patent] is by far the most important recent development in the textile trade, having practically enriched it with a new fibre almost as lustrous as silk, and of course much less costly." Two pages later, in a section on "Cellulose Silk," he noted that "If cotton is mercerized with caustic soda and treated with carbon disulphide while still saturated with the alkali, it forms a new chemical compound . . . known as 'viscose.'" When Gardner wrote these words it was by no means clear which of the many "cellulose silk" processes would ultimately triumph in the marketplace. But long after viscose rayon had come to dominate the cellulosic fibres industry, the step in rayon manufacture where the cellulose is treated with caustic soda was still known as--and probably still is known by some as--mercerizing. Those familiar with this process allude to the mercerizing of cotton textiles as "partial mercerizing."

Space considerations prevent a more extended treatment of the development of viscose rayon in the 1890s and the first two decades of the twentieth century. A significant part of this story has been masterfully told by D. C. Coleman in the second volume of his history of Courtaulds, the British silk manufacturer that through astute technology purchases and much technical efforts of its own pioneered the commercial viscose process. A few points about this history bear noting, however. The first is that the story is really an international one, involving chemists and capitalists from most of the industrialized nations. Second, as the viscose product and process got better during the first two decades of this century, demand for viscose rayon grew and the fibre moved into broader segments of the market. Third, for a firm like Courtaulds and its American subsidiary, American Viscose Company, viscose fibre manufacture was extraordinarily profitable--a situation very different than that experienced by Chardonnet and his backers.

By 1920, the winners and losers in cellulose silk had become apparent. Viscose rayon was king, and nitrocellulose and cuprammonium fibres were on the way down. One other cellulose-based fibre, cellulose acetate, would soon emerge as important though by no means as significant as viscose. Cellulose acetate was principally the handiwork of Swiss chemists, Drs. Camille and Henri Dreyfus, although acetylization of cellulose had commanded the attention of European and American chemists since 1869 when cellulose acetate was first prepared. The work of the Dreyfus brothers is particularly interesting because it emerged from an intensive effort to salvage capital that had been sunk during World War I into a plant to make airplane dope. Whatever the winners and losers, cellulose chemistry in its broadest sense was also king. The intense international interest in artificial fibres and the by no means inconsiderable interest in cellulose chemistry being displayed by the textile industry in the wake of the Lowe mercerizing patent created a situation whereby cellulose chemists were much in demand and highly prized in the 1920s. The Du Pont Company's attempt to hire a seasoned, first-rate cellulose chemist in Europe to head up a new pioneering research laboratory in its rayon and cellophane businesses demonstrates this point quite clearly.

During World War I, Du Pont's executives had committed the company to lessening its dependence on the explosives business. From a multi-year study of investment opportunities in a wide range of chemical businesses (conducted principally by Fin Sparre, a Norwegian-born and German-trained chemist who had directed Du Pont's Experimental Station from 1911 to 1915), the company determined to invest in cellulose fibres. Du Pont was, after all, the world's largest manufacturer of guncotton, and executives thought it obvious and easy to begin manufacturing nitrocellulose fibres. But by 1920 they realized that the future lay with viscose, not with the old Chardonnet silk. In 1921 Du Pont purchased French viscose rayon technology and opened a large plant in Buffalo, New York. Soon the company also purchased French cellophane patents, which also relied upon the viscose process. In 1926 Du Pont also purchased French cellulose acetate technology and entered the manufacture of cellulose acetate fibres. By the end of the decade, the company had penetrated about a third of the U.S. rayon market and a quarter of the cellulose acetate market and, because of its moistureproofing process, had gained what was essentially a monopoly position in cellophane.

With considerable profits pouring in and with huge capital investments in place, the head of Du Pont's fibre and cellophane businesses believed that the company could best protect its profitability and investments in cellulose technology by establishing a Pioneering Research Laboratory. He envisaged the laboratory being headed up by one of the world's finest cellulose chemists who would direct a staff of lesser Ph.D. chemists in pioneering new products and processes. With the approval of his board of directors, this executive sent the director of his existing research organization, which had been working feverishly to rationalize the French technology, to Europe to recruit the new pioneering research director. The Du Pont people thought that they would be able to hire one of the several chemists--mostly Germans--they had identified for about twice the salary of their leading research chemists and directors--for roughly $6000 per year. But to their chagrin, they quickly learned that the chemists who were even moderately interested in moving to the United States wanted not $6000 per year but $6000 per month for their services! Empty-handed, the Du Pont research director returned home and offered the new position to the chemist who had moistureproofed cellophane, William Hale Charch.

For the next thirty years, Charch directed Du Pont's pioneering research in textile fibres, presiding over what can only be described as brilliant industrial research and development. Yet the first twelve years of his tenure as director of pioneering research were not without controversy and angst. This controversy stemmed from Charch's research organization being barred from conducting any research other than in cellulose chemistry. The truly radical development in man-made fibres--synthetics--came from outside Du Pont's cellulosic fibres business and indeed from outside the entire cellulosic fibres industry. Fortunately for Du Pont, the development came from within the company's small central research organization. That development was nylon, the first wholly synthetic fibre, the product that really turned the still moderately sized man-made fibres business into a truly large industry, the polymer that launched the synthetic fibres revolution.

4 NYLON

In 1926 Charles Stine, the director of Du Pont's small central research department who was anxious to build up his organization and the scientific reputation of the entire Du Pont Company, proposed to the firm's Executive Committee that he be allowed to initiate a small, high-caliber program of what he called "pure research." Du Pont's executives politely declined Stine's proposal; evidently they didn't like the concept of "pure" (as in "no strings attached") research. Stine countered with a far more comprehensive and forcefully argued proposal for the establishment in central research of a program of "fundamental research." He stressed that although each of the company's diversified businesses, which were organized into relatively autonomous industrial departments, carried out their own R&D, these organizations (some of which were quite large) were not meeting the long-term research needs of the company. The industrial departments focused their R&D on shorter-term, less theoretical problems. Moreover, Stine argued, even academic chemistry was not addressing problems that were critical to Du Pont's interests, especially problems that cut across the firm's diversified businesses. A first-rate program of fundamental research concentrating on half a dozen or so such areas in chemistry would inevitably benefit the company in tangible ways--in ways that moved beyond his earlier arguments that pure research would improve Du Pont's reputation for scientific research, thereby enhancing the firm's capability to recruit high-caliber chemists and making it easier for the company to obtain research information from other companies and institutions. Stine proposed undertaking fundamental research in colloid chemistry, catalysis, the generation of physical and chemical data, organic synthesis, and polymerization.

Soon after his proposal was approved and generously funded, Stine learned that he could not attract the caliber of chemists he had hoped would fill the positions of group leaders, even though he could offer academic chemists salaries twice as large as those paid by their universities. He was forced to recruit from within his own ranks and from lists of young, aggressive, and very promising academic chemists who were instructors and assistant professors. To lead the research on polymerization, Stine hired an instructor of organic chemistry at Harvard, Wallace H. Carothers, who had never carried out any research on polymerization.

In spite of Carothers' novice status as a polymer chemist, Stine had made a good choice. (This is all the remarkable given the number of fine chemists who turned him down.) Carothers knew that the chemical community had been debating for some time the fundamental nature of polymers. Two predominant views were at odds with one another. One held that polymers were aggregates of peculiar molecules, the structure of which had not been fully explained. The second--and rapidly ascending view--was that polymers merely consisted of long chains of ordinary molecules bound together by ordinary chemical bonds--macromolecules that could be explained using classical chemical concepts. The German chemist Hermann

Staudinger stood as the foremost proponent of this view, but he had not demonstrated his views beyond a shadow of a doubt. Within two years of his joining Du Pont--working in the new Experimental Station laboratory that became known pejoratively as "Purity Hall" by those who did not work there--Carothers and his team of freshly minted Ph.D. organic chemists had published more than a dozen papers including two landmarks, "An Introduction to the General Theory of Condensation Polymers" in Journal of the American Chemical Society and "Polymerization" in Chemical Reviews. In these and other papers, Carothers provided the language that polymer chemists would use thereafter, and he presented a rigorously argued and thoroughly documented case to support Staudinger's views.

Along the way--and certainly not planned--Carothers and his group polymerized neoprene synthetic rubber (called initially "chloroprene," thereby drawing the analogy with isoprene) and the world's first wholly synthetic fibre (an aliphatic polyester). These discoveries came within two weeks of each other. Du Pont's Organic Chemicals Department quickly picked up the development of neoprene, owing to its increasingly strong position in manufacturing chemicals for the rubber industry and the considerable scientific and technical expertise of chemists in its Jackson Laboratory. But importantly, the central research unit retained total control of the synthetic fibre work. The reason was quite simple. Stine's new assistant, the same research director from Du Pont's rayon business who had chosen Hale Charch to direct pioneering research after failing to hire a top-caliber German cellulose chemist, knew that in spite of Charch's group's achievements in cellulose chemistry, it possessed neither polymer synthesis capabilities nor permission by the top management of the rayon business to pursue any research other than in cellulose chemistry. Indeed this short leash had been one of the factors precipitating his transfer from the rayon business to central research.

Carothers and his assistants were certainly aware in an elementary way about what properties were necessary to make a synthetic fibre with commercial potential. Naturally, they had been paying particular attention to how far they could drive up the molecular weight of some of their synthetic polymers, and they had a tacit belief that a polymer with a molecular weight somewhere between 10,000 and 12,000 might be spinnable as a filament. Then, of course, they knew about such properties as melting point; stability in light, solvents, and water; and, in a rough way, desirable aesthetic properties. After Carothers and his group managed to polymerize the spinnable (and cold-drawable) 3-16 polyester in March 1930, they worked to find a polymer with a higher melting point, greater resistance to hydrolysis, and better physical properties. But they failed. Carothers remained largely unconcerned because most of his theories had been borne out, and he was ready to undertake research in large ring structures. Besides, he had not been hired to invent a synthetic fibre; he was supposed to do first-rate fundamental research and publish his results.

But events within Du Pont--Stine's promotion into the top management of the company and his succession by a more industrially oriented, hard-boiled chemist (Elmer K. Bolton) who had opposed Stine's fundamental research program in the first place--and developments outside the company--the deepening economic depression--brought about a major change of climate in Purity Hall. After some stormy and controversial times in which it had become apparent that the fundamental research program was expected to pay its way, Carothers seems to have consented to take up again the question of synthesizing a polymer that could be spun into a commercially promising fibre. He had finished his research on large ring structures, and apparently he was unsure of the next step he should take.

Early in 1934 Carothers outlined a new attack on the problem of obtaining a good synthetic fibre. Yet his approach would still involve a class of polymers that he and his assistants had tried earlier--the polyamides. Carothers had turned to the polyamides in 1930 because his theories of condensation polymers and his sense of structure/property relationships suggested they would have a higher melting point and be more stable in water. But he and his group members had been unable to obtain a tractable polyamide with a high enough molecular weight. Now Carothers laid out a polymerization route to a commercially viable fibre starting from an aminononanoic ester monomer. One of his group members spent five weeks preparing this compound. Once he had obtained enough of it, he easily polymerized it. On May 24, 1934, he melted the polymer in a bath at about 200 degrees C, put a stirring rod into the molten polymer, and pulled out a fine filament that could be cold drawn into a beautifully lustrous filament. The filament proved to have properties that immediately suggested its commercial potential. Its tensile strength was outstanding, and its wet strength was unmatched. Moreover, the silk-like fibres were highly abrasion resistant.

From this point, much of the central research unit, not just Carothers's research group, moved into a campaign mode to develop what became nylon fibres. All notions about fundamental or pure research were cast aside. The dozens of potential acid-amines pairs of the polyamides had to be synthesized and their properties checked. The necessary intermediates for these different syntheses had to be prepared. Potential commercial synthesis routes for intermediates and monomers alike had to be explored. Spinning methods had to be considered.

The history of the brilliant development of nylon is beyond the scope of this paper. But two points are significant. First, Du Pont's rayon R&D team was not brought into nylon's development until well into the project. This team contributed much more to the plant process and product development than to any work in synthetic polymer chemistry. Only after the commercialization of nylon in 1940 did the rayon research organization build organizational capability in synthetic polymer chemistry. Second, the introduction of nylon at the 1939 New York World's Fair--fifty years after Chardonnet silk debuted at the Paris World's Fair--proved to be sensational. As Charles Stine touted,

> "This textile fibre is the first man-made organic textile fibre prepared wholly from new materials from the mineral kingdom. I refer to the fibre produced from nylon. . . . Though wholly fabricated from such common raw materials as coal, water, and air, nylon can be fashioned into filaments as strong as steel, as fine as a spider's web, yet more elastic than any of the common natural fibres."

Simply put, nylon was the first of the "miracle fibres." Initially targeted to replace expensive silk in women's hosiery by nylon, the new fibre began to turn a profit while it was still in the pilot plant stage in 1939. No fibre subsequently developed by Du Pont would do that, and none would ever be as commercially successful as nylon. It was Du Pont's and the world's first billion-pound synthetic textile fibre, and it has generated profits for the company of perhaps $20 billion. When the U.S. government diverted all of Du Pont's new nylon capacity to the war effort--for parachutes and bomber tire cords, for example--women were deprived of the stockings that had only briefly graced their legs before the war. When the war was over, women in cities across the United States stood in lines that were often several blocks long to buy a pair of "nylons." Riots even broke out, with women fighting one another to get their hosiery.

5 THE SYNTHETIC FIBRES REVOLUTION

Nylon took the rayon and cellulose acetate manufacturers by storm. And it turned plenty of heads in the textile industry as well. Cellulosic fibres manufacturers throughout the world moved as quickly as they could to undertake research in synthetic fibres, as did chemical firms and the nascent petrochemical companies. Their chemists poured over Carothers's patents, which had begun to be issued in 1937. Du Pont itself, in a grave error, informed its sometimes ally and sometimes enemy I.G. Farben about its nylon project in 1937. The German chemists went over every inch of Carothers's work and found a fundamental weakness in the impressive patent position that Du Pont had built. In an early paper Carothers had stated categorically that caprolactam could not be polymerized either with or without a catalyst. The Germans proved him wrong and in the process developed caprolactam nylon or nylon 6.

Two research chemists at the Calico Printers Association, Ltd., in Lancashire, J.R. Whinfield and J.T. Dickson, also pondered Carothers's work. In 1940, the year that nylon was commercialized, they followed a polymerization process spelled out in great detail by Carothers, but they substituted terephthalic acid for the aliphatic acids used by Carothers and achieved a non-hydrolyzing, fibre-forming polyester with a high melting point and outstanding tensile properties. Although Du Pont's Hale Charch initially argued that Whinfield and Dickson's work was dominated by Carothers's basic patents, he soon concluded that Du Pont had to purchase the exclusive rights to their American patents. This recommendation was wise, and Du Pont moved more aggressively that the British owners of the Whinfield and Dickson patents to develop polyester into an important textile fibre, introducing Dacron polyester fibre commercially in 1952. Indeed, Dacron polyester became Du Pont's second billion-pound fibre and actually surpassed nylon in tonnage (but by no means in profitability) in 1973.

Du Pont's development of Dacron polyester closely paralleled its development of the firm's second major synthetic fibre, Orlon acrylic, which was commercialized in 1950, two years before Dacron polyester. In both these fibre development projects, natural fibres--especially wool--figured very substantially. Well before Hale Charch had been given the mandate in 1940 to build an outstanding polymer synthesis capability within his Pioneering Research organization, he had assigned some of his best researchers to try to make a wool-like rayon. During the late 1920s and throughout the 1930s, the rayon industry had received a boost when the textile manufacturers to whom they sold their fibres had manufactured rayon cloth from staple yarns (as opposed to filament yarns). Although cloth made from delustered rayon looked like wool, it possessed none of the feel, or handle,of woollen cloth, nor did it exhibit any of wool's resiliency. Charch's researchers were, therefore, trying to build resiliency into rayon by cross-linking its molecules with synthetic polymers. The very idea of cross-linking molecules within fibres stemmed from research being done in natural fibres, especially the work carried out in England at Tootal Broadhurst Lee in self-smoothing or no-iron fabrics--research that fell within the tradition of textile chemistry initiated by John Mercer.

One of Charch's chemists assigned to the rayon cross-linking project was supposed to be trying to employ the long-known, highly intractable polyacrylonitrile as the cross-linking synthetic by polymerizing it in situ in the rayon filaments. Failing in his experiments, this chemist determined in 1941 to find a solvent to dissolve the polymer and then spin it into a fibre all by itself. After an extensive search for a solvent, which was aided by a suggestion from one of Du Pont's principal academic consultants, the chemist succeeded in finding one and in spinning a highly attractive acrylic fibre. If left undrawn or only slightly drawn and then crimped, the fibre seemed to have wool-like properties. If drawn substantially (6x), the fibre took on a lustrous, silk-like appearance.

Convinced that the most logical market to aim for in a post-nylon synthetic fibre was wool, Charch spent the rest of the decade of the 1940s preaching a gospel of wool substitutes. His laboratory gathered an impressive array of different wools from all over the world and read everything that was published on the chemistry and physics of wool. Initially, Charch had been convinced that to get a good synthetic wool, one would have to create what he called "a Chinese copy" of the natural fibre--in other words a crimped fibre complete with scales and all. But Pioneering's best physical chemists abstracted the essence of wool--its resiliency--by developing a rigorously analytic concept of resilience and then constructing a three-dimensional map of fibre resilience. By focusing on the property of resiliency, Pioneering researchers were able to evaluate wooliness in synthetic fibres by comparing their position on the resiliency map with real wools. And by carrying out research work on polymer structure/polymer property correlations, Pioneering's researchers were able to enhance resiliency in both Orlon acrylic and especially in Dacron polyester. Owing much to Charch's commitment to developing wool substitutes, both Orlon and Dacron rapidly penetrated the market for wool, initiating its decline. Moreover, by literally building resiliency into Dacron polyester, once polyester became cheap enough, it could be blended with cotton to yield a remarkably strong, no-iron fabric that has since the mid-1950s come to dominate the cotton shirting markets. Research on natural fibres, done both within Du Pont and outside in numerous research institutes, universities, and companies, thus played an important role in the innovation of the fibres that drove a good percentage of those fibres out of the market.

By the late 1940s, Hale Charch was so confident in his organization's capabilities in synthesis, spinning, structure/property analysis, and the like, that he claimed that Du Pont was on the verge of being able literally to engineer fibres to meet any desired fibre specification. He announced that he and his staff were but at the beginning of a "synthetic fibres revolution." He predicted that nylon, Orlon, and Dacron would deeply penetrate natural fibres markets and that it would be difficult to develop another really big-tonnage synthetic fibre. Because of this, he suggested that his laboratory should devote its efforts to the development of specialty fibres--fibres that, for instance, would have elastomeric properties superior to fibres made from natural rubber; fibres that would possess tensile strengths beyond anything yet attained with synthetics and anything imaginable with natural fibres; and fibres that would be flame resistant and that would perform at extremely high temperatures.

In most respects, Charch's predictions were accurate. Indeed, he and his contemporaries were at the beginning of a synthetic fibres revolution. Synthetic fibres have deeply penetrated natural fibres markets, and aside from polypropylene, whose significance can be debated, another big-tonnage fibre has not been developed. Specialty fibres have become an important part of the synthetic fibres industry, and it is in this area that chemistry has come a long way. One need look no further than the amount of chemical research that has been done in the wake of the synthesis of the first liquid crystalline polymers, polymers that eventually appeared in fibre form in products like Du Pont's Kevlar aramid fibre and Akzo's aramid product. But in some respects Charch failed to anticipate how resilient the natural fibres industry would be in the face of severe competition from synthetics. Nor did he anticipate that there might be a backlash against apparel made from synthetics. Without better performing natural fibres, which owe their enhanced properties to textile chemistry, the backlash might not have been as severe.

6 THE IMPROVEMENT OF NATURAL FIBRES

In some respects, we have come full circle--back to the implications of John Mercer's work, back to the idea that natural fibres can be improved through chemical modification. As Sherman and Lidfeldt wrote in 1946, "Chemical treatments are the cotton and wool industries' answer to the threat of the new fibres. Only by offering better cotton and wool can these industries hope to keep up with the pace of such new competition." The research on the chemical modification of natural fibres carried out by industry- and government-sponsored textile institutes in the United States and Britain during much of the twentieth century has yielded impressive results.

The development of durable-press cotton alone provides an important chapter in the history of textile fibres. Durable-press techniques faced considerable obstacles, including resistance from the textile trades and the widely varying laundering habits of textile consumers. Fishy odors from self-smoothing resin treatments had to be overcome; increased tendency to soil and stain had to be eliminated. Blending of cotton and polyester posed its own set of problems and drove much of the research. Moreover, the hydrophobic nature of polyester, acrylic, and nylon fibres posed severe problems in dyeing them, which were only heightened with blending the synthetics with natural fibres. Extensive fundamental research in dye chemistry, addition of molecules to the homopolymer to enhance dye affinities, and development of entirely new dye processes were required before these problems were solved.

Research done in the chemical modification of wool during this century to my mind is all the more impressive in spite of the decline of its importance to consumers on a per capita basis. In 1870 Americans consumed almost five pounds of wool per year per person. This figure remained almost constant, peaking in 1946 at about 5.6 pounds per capita. Under assault from the synthetics, wool consumption fell to 1.2 pounds per capita in 1970, a year that saw synthetic fibers approaching their zenith on a U.S. per capital consumption basis. Wool is a far more complex fibre than cotton, both physically and chemically. Understanding it on a fundamental basis has required knowledge going far beyond conventional textile chemistry. W. T. Asbury, John B. Speakman, and their team at Leeds University must be viewed as the real pioneers in the fundamental study of wool, with their research beginning in the late 1920s and early 1930s. A complex protein, wool is composed of some twenty different amino acids. Understanding it, therefore, demanded knowledge of biochemistry--knowledge most often associated with the likes of Linus Pauling, Frederick Sanger, James Watson, and Francis Crick.

Determining the precise amino acid sequencing in protein molecules has been intellectually and instrumentally challenging. Two chemists within the British Wool Industries Research Association, Archer J. P. Martin and Richard M. Synge, rose to that challenge and developed partition chromatography in 1941. For their contributions to basic chemistry, they were awarded the Nobel Prize in Chemistry in 1952.

Not all chemical research in the wool industry yielded Nobel Prizes, but it certainly has led to significant developments in wool itself. Chemical modification of the surface of wool resulted in shrink-proofing it, yielding fabrics that can be cared for more easily by consumers accustomed to automatic washing machines. Moth-proofing has also been achieved, and luster has been added to wool, providing it with greater range as a textile fibre. Cross-linking and in-situ interfacial and solution polymerization have been employed in the chemical modification of wool.

Since World War II, chemistry has played an increasingly important role in the textile industry. Flame-retardants, stain-retardants, and static reducers have commanded a tremendous amount of attention by textile chemists. In some instances, new chemistry has been explored. But the history has not been one of continuous, smooth-sailing. Indeed, the greater role of chemistry in the textile industry has introduced problems that formerly did not exist to the same degree. One example clearly makes this point.

7 SOME UNANTICIPATED PROBLEMS

Beginning in the late 1960s in the United States, consumer product safety advocates began to voice concern about the safety of children's sleepwear. Both highly brushed cotton flannels and the soft and warm-feeling acrylic and modacrylic garments were implicated in the 300 infant deaths and 6000 serious injuries annually attributed to flammable infant sleepwear. By 1973, the federal government had imposed regulations banning the manufacture of children's sleepwear that was flammable according to the standards set by the Commerce Department. These regulations would be enforced by the Federal Consumer Product Safety Commission. For manufacturers, the solution was already at hand: treat sleepwear with the flame retardant tris(2,3-dibromopropyl) phosphate (Tris).

Three years later, however, a major controversy

erupted about the safety of the flame retardant. In 1975, someone put Tris through the test for carcinogenicity that Bruce Ames had developed the year before and found the results strongly positive. The Consumer Product Safety Commission took no immediate action. By April 1976 controversy over Tris's safety had led to its being banned, and there was widespread fear among Americans that their babies were being poisoned with Tris. The Consumer Product Safety Commission, which suffered a major black eye in the incident, was forced to issue a recall of all Tris-treated garments, the total bill for which added up to some $200 million, borne after the usual legal disputes by the chemical manufacturers, the textile makers, and the apparel makers. More alarming for consumers were reports that other fibre treatments for flame retardation, shrink-proofing, moth-proofing, and the like might also be carcinogenic. Already in a recession, the textile industry and the fibre industries encountered hard times. Textile chemistry faced the same problems and challenges that other segments of the chemical industry were also experiencing, that of operating in an environment of greater regulation with increasing concerns over carcinogenicity, mutagenicity, and pollution.

Synthetic fibres manufacturers experienced a perhaps more disturbing reality during the 1970s. The OPEC-induced oil shocks of that decade demonstrated to them that the synthetic fibres revolution had been not just a product of impressive developments in polymer chemistry and chemical engineering, nor had it been merely a result of important shift in consumer preferences toward emphasis on utility and ease of care in apparel fabrics. The synthetic fibres revolution had also been a product of an era of relatively cheap energy. In the face of a crisis in the supply of energy, some industry experts wondered what the future held for synthetic fibres. In spite of the apparent abundance of oil in the 1980s and early 1990s, that question is still one to be pondered.

BIBLIOGRAPHY

Avram, Mois H. The Rayon Industry. (New York: D. Van Nostrand, 1927).

Barker, A. B. et al. Textiles. (New York: D. Van Nostrand Company, 1910).

Bud, Robert and Gerrylynn K. Roberts. Science vs. Practice: Chemistry in Victorian Britain. (Manchester: Manchester University Press, 1984).

Chemistry in the American Economy. (Washington, D.C.: American Chemical Society, 1973).

Clow, Archibald and Nan L. Clow. The Chemical Revolution. (London: Batchworth Press, 1952).

Coleman, D.C. Courtaulds: An Economic and Social History. Vol. 2, Rayon. (Oxford: Clarendon Press, 1969).

Dictionary of National Biography. s.v., "Mercer, John."

Haynes, Williams. Cellulose: The Chemical That Grows. (Garden City, NY: Doubleday, 1953).

Hartsuch, Bruce E. Introduction to Textile Chemistry. (New York: John Wiley & Sons, 1950).

Hounshell, David A. and John Kenly Smith, Jr. Science and Corporate Strategy: Du Pont R&D, 1902-1980. (New York: Cambridge University Press, 1988).

Linton, George F. Natural and Manmade Textile Fibers. (New York: Duell, Sloan & Pearce, 1966).

Markham. Jesse. Competition in the Rayon Industry. (Cambridge, MA: Harvard University Press, 1952).

Marsh, J.T. Self-Smoothing Fabrics. (Metuchen, NJ: Textile Book Service, 1962).

Mauersberger, Herbert R. American Handbook of Synthetic Textiles. (New York: Textile Book Publishers, Inc., 1952).

Moncrieff, R. W. Man-Made Fibres. (New York: John Wiley & Sons, 1978).

Moncrieff, R. W. Wool Shrinkage and its Prevention. (New York: Chemical Publishing Co., 1954).

New York Times.

Parnell, Edward A. The Life and Labours of John Mercer, F.R.S., F.C.S. (London: Longmans, Green, & Company, 1886).

Sherman, Joseph V. and Signe Lidfeldt. The New Fibers. (New York: D. Van Nostrand Co., 1946).

Synthetic Dyestuffs: Modern Colours for the Modern World

Anthony S. Travis

SIDNEY M. EDELSTEIN CENTER FOR THE HISTORY AND PHILOSOPHY OF SCIENCE, TECHNOLOGY AND MEDICINE, THE HEBREW UNIVERSITY OF JERUSALEM, GIVAT RAM CAMPUS, JERUSALEM 91 904, ISRAEL

Introduction

Between 1865 and 1875, as artists began to manipulate nature on canvas in ways that gave rise to the style later called Impressionism, chemists began to manipulate nature on their benches in ways that would become the basis of modern technological society. The attentions of these artists and chemists converged with the novel use and creation of colour, and quite often their interests were directed at the same subject, the fashion world.

It hardly requires a visit to a major art museum, auction room, or bookstore, to appreciate the impact of Impressionism on our culture and lives. But what of the impact of the chemists who gave us new colours? Do we recall the names of Perkin, Caro, and Baeyer as readily, and with the same emotions, as Monet, Renoir and Degas? And is the artist Max Liebermann better known than another member of his family, Carl Liebermann, who teamed up with Carl Graebe at the end of the 1860s to replicate nature in the laboratory when they synthesized the red dye alizarin? These are, of course, rhetorical questions for most people. If the names of great colour chemists cannot be allowed to rank with those of great artists, the 150th anniversary of the Chemical Society of London at least provides an appropriate opportunity to review a century and a half of their remarkable creative activities.

The synthetic dyestuffs industry began in Britain with the discovery by William Perkin of mauve in 1856. This was made from aniline, produced from coal tar, the waste of gas works. Because the colour was so well suited to silk, the industry grew rapidly in Lyon, France, where red, blue and green dyes were also obtained from coal tar derivatives.

During the 1860s, several German and Swiss dye-making companies were formed to exploit the new inventions that served the increasingly mechanised European textile industry. They are familiar to us as BASF, Bayer, Hoechst, AGFA, CIBA and Geigy. In the 1870s and early 1900s scientific and technical advances enabled synthetic alizarin and indigo to oust the natural products from the madder root and the indigo leaf, respectively. These, and new dyes with no analogues in nature, were also made from coal tar, and paved the way to pharmaceutical and photographic products, and explosives, for mining, railway construction...and warfare.

Early in the 20th century the German and Swiss dye industries dominated the field. The First World War led to the revival of the British industry, and the creation of the American dye industry. From the 1920s, demand for improved colour fastness and for dyes that could be attached to synthetic fibres provided new challenges. British industrial chemists were at the forefront of developments that eventually led to fibre reactive and disperse dyes, associated with ICI.

The Industrial Revolution and Demand for Dyestuffs

The Chemical Society of London was formed a decade before the Great Exhibition in London. Iron machines and steam power had transformed society, and the most far reaching changes had taken place in the textile industry. Greater output called for new and faster methods of bleaching and processing cloth and fibres. Demand for dyestuffs, like madder and indigo, rose in line with increased production, especially of cotton goods. The major manufacturing centres were Lancashire and south-west Scotland, and chemists had been attracted to these areas from the late 18th century onwards. Their successes with bleaching, extracting colours, and fixing them, were appreciated by dyers and calico printers, who were soon themselves studying elementary chemistry in order to keep up with, and implement, the latest advances.

By 1840, the influence of chemistry on textile dyeing and printing was so great that textbooks on the subject began to appear. "Colouring matters" were frequently investigated by chemists, like F. F. Runge in Germany who found several colour reactions from coal tar products. Justus Liebig and Friedrich Wöhler studied the steps leading to the colour murexide, a purple that was later compared with the famous dyestuff of antiquity. However, few of these novel substances found application in dyeing and printing, especially as the methods of fixing new dyes to natural fibres often created considerable technical difficulties.

The major change in this state of affairs occurred after 1845,when an institution with close connections to the Chemical Society, the Royal College of Chemistry in London, first opened its doors. The director was August Wilhelm Hofmann, a former assistant of Liebig in Giessen and an expert on coal tar compounds, especially aniline. Hofmann realised that the college had to provide skills for those engaged in the practical application of chemistry, and also to undertake research that might lead to useful products, especially new drugs and dyes. Coal tar provided a variety of hydrocarbons, like benzene, naphthalene and anthracene, that formed the basis of Hofmann's investigations. Their amino compounds were of special interest because they appeared to be related to the alkaloids, that included natural drugs like quinine, and because aniline, the simplest aromatic amine, was well known for its colour reactions.

Hofmann's research efforts produced much of academic interest, but no new useful compounds. Funding by sponsors interested in discoveries that could be applied to industry and agriculture was on the wane by 1850, and the government stepped in to save the Royal College of Chemistry. It became part of the Royal

School of Mines in 1853, but retained its separate identity. If the fortunes of the college did not improve, then at least its image soon did. For in the same year Hofmann accepted the fifteen-year-old William Henry Perkin onto the teaching course. Perkin's enthusiasm for chemistry was soon rewarded with a research project aimed at making an amino compound from anthracene. Although not successful, the effort convinced Hofmann of his teenage student's dexterity. Perkin was guided towards new and more promising experiments on amino compounds, in collaboration with his friend Arthur Church, and coloured products were carefully tested for their dyeing properties.

The breakthrough occurred during the Easter holiday of 1856. Inspired by Hofmann's research programme, Perkin attempted to make quinine by oxidative condensation of allyltoluidine. This was undertaken in a primitive laboratory he had fitted up in his parents home, located in the East End of London. The experiment failed, but Perkin decided that further investigations might not be amiss and repeated the reaction on aniline, using the same oxidant, potassium dichromate. This afforded an unpromising black precipitate, but treatment with alcohol gave an intriguing result: a purple solution of remarkable intensity. The colour imparted colour to cloth, especially silk, and resisted washing and exposure to light. Perkin had, almost by chance, discovered the first of the aniline or coal tar dyes. His chemical training, and involvement with the amino compounds from coal tar hydrocarbons, had been critical elements of his achievement.

<u>George Perkin & Sons</u>

By the end of 1857, William Perkin, in partnership with his brother Thomas, and their father, George, had opened a small factory at Greenford Green, north-west of London, for manufacture of the new colour, at first called Tyrian Purple, after the city of Tyre where the ancient purple was made. In the factory they installed glass apparatus for small scale production of the intermediates, nitrobenzene and aniline, made from benzene, and plant for the oxidation step and subsequent extraction of the dye. By this time the guano-derived murexide colour was also available commercially, as was another coal tar derivative, the yellow picric acid, or trinitrophenol. In addition, a dyer in Lyon had produced from lichen a fast purple similar to the aniline colour. Purple was becoming increasingly fashionable, and the new products were admired by printers and dyers, as well as by the ladies in the high street. It seemed an opportune moment for George Perkin & Sons.

However, Perkin's new colour came up against a major obstacle. It was chemically basic, in contrast to the acidic plant derived dyes, and thus the traditional ways of fixing were of no use, especially to cotton, the basis of bulk production. As a consequence, dyers and printers were not prepared to introduce the colour. This presented a new challenge to Perkin, one that he tackled with the same single-minded determination that he had applied to the development of his aniline purple. Working closely with a dyer and printer in Scotland, Perkin found mordants which saw sufficient use to enable quite widespread adoption of the purple. Perkin's manufacturing process was copied by companies in Lyon and Paris during 1858-59. Since the aniline and the new lichen purples were the leading fashion colours of Paris, it was only a matter of time before the craze moved to London and other European cities. In May 1859, the aniline purple, by then known throughout Britain as mauve, was joined by an eye-catching brilliant red, called fuchsine by the first producers, Renard Frères of Lyon.

William Perkin refused to license his patented process for mauve to other British chemical manufacturers. One of these, Simpson, Maule & Nicholson, of London, found a lucrative market for the intermediates, and enjoyed a good export business for aniline. Edward Chambers Nicholson, a partner, also found an improved method for the red, using arsenic acid for the oxidative condensation of commercial aniline. This was discovered at about the same time by Henry Medlock, who filed a patent a week before Nicholson. Medlock's patent was soon in the hands of Simpson, Maule & Nicholson, which became the largest producer of the red in Britain, where it was known as magenta, after the battle of that name, or roseine.

Simpson, Maule, Nicholson and Medlock were, like William Perkin, former students of Hofmann. The latter's high reputation drew a number of German students to the Royal College of Chemistry, notably Peter Griess and Carl A. Martius. Griess moved to a brewery in Burton on Trent, and Martius took on a post at Roberts, Dale & Co. in Manchester. Martius was encouraged to join this firm, and Griess was brought in as consultant, by Heinrich Caro, one of several young German chemists and colourists who had been attracted directly to Manchester, the hub of the world's largest textile industry, and hence the centre for dye production and use.

New Aniline Dyes

Simpson and partners expanded greatly, assisted by the scientific services of Hofmann, and by the introduction of new dyes, such as aniline blue. This was discovered early in 1860 by Charles Girard and Georges de Laire working in a French owned chemical factory at Brentford, near London. They studied the new blue with Hofmann, and passed over the patent rights to the largest French manufacturer, Renard Frères. The blue was licensed by Renards to Simpson's firm. In May 1863, Hofmann showed that the blue was phenylated magenta, in other words a substitution product. He immediately tried other substitutions, using alkyl iodides, and obtained a brilliant violet. This was the first synthetic dye that resulted from well reasoned prediction based on chemical research. The commercial product was known as Hofmann's violet. Through this industry-related work Hofmann was enabled to investigate the chemistry of aniline dyes, and showed that the red was formed because the amino compound from toluene, toluidine, was present in the commercial aniline. This helped manufacturers to obtain greater control over the reaction.

aniline + toluidine = aniline red (magenta, fuchsine)

aniline red + aniline = aniline blue

aniline red + alkyl iodide = aniline violet (Hofmann's violet)

Figure The production of aniline red and its derivatives.

Simpson, Maule & Nicholson and Renards guarded their respective aniline red patent monopolies jealously, and took strong actions against supposed infringers. This encouraged innovations elsewhere, especially of processes that did not involve oxidations on aniline or toluidine. The results were phenol-based colours, naphthalene colours, modifications of mauve, and exploitation of Griess' discovery, the diazo reaction. Roberts, Dale & Co., under Caro's guidance, developed all these processes, and borrowed new methods and products from Germany.

Thomas Holliday, son of the tar distiller Read Holliday of Huddersfield, liberated the aniline red process from patent monopoly in 1865. This led to a rapid restructuring of the industry in Britain. One of the newcomers was Ivan Levinstein, from Berlin, who set up a works near Blackley, Manchester, later to become a principal component of ICI.

In France, by contrast, Renards retained exclusive control over the production of aniline red, blue and violet, and their main oxidant used to manufacture the red after 1860 was arsenic acid. From 1866, industrial chemists in Paris and Lyon scaled up a useful new reaction, one in which amino group hydrogens of aniline and toluidine were replaced by phenyl and alkyl groups prior to oxidative condensation. This overcame the problem of arsenic-poisoned wells and rivers, since a different oxidant could be used, and it also enabled certain patented processes to be successfully circumvented.

Sample of Heinrich Caro's printing trials using mauve and aniline black made by Roberts, Dale & Co., Manchester, 1862. (Courtesy of Deutsches Museum, Munich).

The method was rapidly adopted in Britain, where Levinstein's Blackley blue was one of the most successful products resulting from this new mode of making coal tar dyes. Gradually the range of synthetic dyes was enlarged to encompass not only various shades of red, blue, and violet, but also greens, yellows, brown, and grey. There was also a popular black that Roberts, Dale & Co. had isolated from the mauve reaction mixture. The aniline dyes supplemented, and often displaced, plant and mineral dyes.

The mid 1860s saw the return from Britain to their homeland of many of the German visitors. Hofmann departed for Berlin in 1865, Caro left Manchester for Heidelberg in 1866, and Martius returned to soon become a cofounder of AGFA. Caro worked with Bunsen, and as a private consultant mainly for the Badische Anilin- und Soda-Fabrik (BASF), formed in 1865. He met the partners earlier in the 1860s as a result of his work at Roberts, Dale & Co. Late in 1868, Caro was taken on by the BASF to develop the novel dyes he had invented in Manchester, including the blue indulin, and to investigate the commercial possibilities of a new process that would transform the fledgling dye industry. This was the artificial alizarin process of Carl Graebe and Carl Liebermann.

Alizarin

Graebe and Liebermann were students of Adolf Baeyer at the Gewerbe Akademie, Berlin. Baeyer's position in this leading trade school meant that, like Hofmann, his research was often guided by utilitarian objectives. The production of benzoic acid for the aniline blue process, and studies on the nature of indigo, were topics of great interest to dye makers and users, and were studied by Baeyer and his assistants. Kekulé's benzene ring theory of 1865 enabled a new understanding of the nature of the coal tar - or aromatic - compounds, that was eagerly followed up in Germany. Emil Erlenmeyer in Heidelberg first drew naphthalene as two fused rings in 1866. This hydrocarbon was thought to be the "mother substance" of the dye from madder, alizarin, but Graebe and Liebermann found otherwise. By skillful degradation and analysis, they showed that alizarin was a dihydroxyquinone derivative of anthracene. Despite incomplete structural knowledge, the available information was sufficient to attack the synthesis. Within months, they managed to prepare alizarin from anthracene. The method gave a low yield, and required the expensive element bromine. Nevertheless, patents were filed at the end of 1868, and purchased by BASF early in 1869.

Heinrich Caro was given the task of finding a more suitable route to alizarin, of procuring adequate supplies of the rare anthracene, and of scaling up the process. He achieved a viable route to industrial synthesis by showing that the introduction of hydroxyl groups could be brought about smoothly using a slight modification of a new reaction: sulphonation followed by hydrolysis. At the same time, in May 1869, the reaction sequence was discovered independently by the Hoechst dyeworks, and, at Greenford Green, by William Perkin.

Perkin patented his process in London on 26 June 1869, one day later than Caro, Graebe and Liebermann. It was a difficult situation, and neither side wished to become involved in litigation. Also, there was the threat of other companies to consider, especially Roberts, Dale & Co. whose consultant Carl Schorlemmer, at

Owens College, was working on an artificial alizarin process. From November 1869, Caro and Perkin began negotiations which led to cross-licensing arrangements and to the sharing of knowledge. Perkin was the first to successfully produce alizarin on an industrial scale and he greatly assisted BASF with technical development. It was another British success story, but, for nearly fifty years, one of the last in the field of synthetic dyestuffs.

German Advances

Synthetic alizarin was an overwhelming success, since the colour could be applied without the development of new processes. The natural product, madder, was displaced during the 1870s, as was the leading position of the British dye industry. After the Franco-Prussian War of 1870-71, it was the German dye industry that rose to first place. German marketing methods were far superior, and German academic chemistry was much better attuned to the needs of the dye industry. This did not go unnoticed in Britain, and there were several moves towards the improvement of technical education and the training of chemists.

The Germans also mastered industrial-academic collaboration, which became a model for modern science-based technological industries. Caro's meetings and correspondence with Baeyer led to the discovery of the fluoroscein-derived dye eosin in 1874. This was the first of the phthalein colours, made by reactions between phthalic anhydride and phenols. Britain was a source of phthalic anhydride, manufactured from the hydrocarbon naphthalene, and remained the major supplier of coal tar hydrocarbons, and many important intermediates, until the end of the 19th century.

The Development of Azo Dyes

The Swiss-trained Otto N. Witt was the pioneer of successful azo dyes, the first of which was introduced in 1875 by his English employers, Williams, Thomas & Dower, at Brentford. To achieve this, Witt investigated the diazo reaction, used to afford strongly coloured products insoluble in water. In 1866, Kekulé had made a notable advance when he proposed the diazo radical, in which two nitrogens replaced one hydrogen of the benzene ring. For the next decade chemists pondered on the puzzle of the constitutions of the coloured substances. Witt searched for an unknown compound of this type containing two free amino groups, and believed to be the missing link in a series that included known yellow and brown dyes introduced in the early 1860s at Roberts, Dale & Co.

Witt speculated that the compound was an orange colour, and also that it would be of moderate stability. He was drawn to the work of Baeyer and Jäger who had suggested the notion of coupling of groups of atoms in the reaction leading to what are now called azo dyes. Witt soon discovered the desired dye, with the predicted properties, and it was marketed as London yellow. He shared details of his process with Caro and Griess, and they agreed that the information should be kept secret. In France, François Roussin made azo compounds based on naphthalene, taken up by the Poirrier firm in Paris.

Martius, now a partner in AGFA of Berlin, acquired a sample of the BASF version of the new dye, named chrysoidine, and handed it over to his consultant,

Hofmann, who quickly published the formula. The colour reaction fell into the public domain, and numerous firms throughout Europe exploited the process, to the great disappointment of Witt and Caro. The azo field was brought under control in Germany during 1877, following the passage of a comprehensive patent law. This encouraged the organisation of industrial research laboratories, dedicated to the discovery of new azo colours and other colour-rendering reactions.

Colour and Constitution

Chemists used the diazo reaction to produce compounds of the type Y-N=N-X, in which different combinations of X and Y were introduced. This involved thousands of experiments, most of which gave coloured compounds that were of no use as dyes. Nevertheless, a degree of theory-based design of colours was possible. For azo dyes, wide-ranging structural variations relied on the availability of intermediates created during the first two decades of the dye industry. Witt, for example, added resorcinol to a diazo intermediate to afford an orange dye. Other azo dyes were made from the reaction between intermediates and compounds in which amino group hydrogens had been substituted. When toluene was used instead of benzene redder shades were obtained.

Witt's 1876 theory concerning colour and molecular constitution was a milestone in the understanding of azo and other industrially important colours. The theory concerned the known groupings of chemical elements that appeared to be necessary for the quite separate and specific properties of colour and dyeing. By specifying the nature of these groups, it was possible to narrow down the range of compounds that might be both coloured and capable of adhering to fibres. There were several refinements of Witt's original theory, including Henry E. Armstrong's quinonoid theory of 1888.

Competition and Cartels

Several azo dyes were developed by the British dye industry, especially at the factories of Levinstein and Read Holliday. This also brought them into patent litigation with their German rivals. At the same time, German dyemakers considered the possibility of buying out British firms, partly in response to anticipated changes in the patent law, but also to retain a secure foothold in the world's largest dye consuming country. For a time Bayer and AGFA had interests in Levinstein's factory, and used him to maintain the strength of a cartel arrangement called the benzopurpurine convention. The benzopurpurines were also azo dyes, and derived their great importance from the fact that they could be used to dye cotton without the aid of a mordant. The first one was discovered in 1884 and marketed by AGFA as Congo red. It was the Bayer company that introduced the version known as benzopurpurine, and which joined with AGFA to create the convention. Levinstein, on their behalf, defended the convention in legal actions against Dawson Brothers and Read Holliday, both of Huddersfield. Once freed from the control of German partners, Levinstein fought for patent reform, especially the necessity of working foreign patents in Britain. This was achieved in 1907, but was not enough to revive the British dyestuffs industry.

Another direct dye, primuline, was created by Arthur Green during 1887 at Brooke, Simpson & Spiller, successors to Nicholson's firm. This was commercialized by Green in the 1890s at Clayton Aniline Co., Manchester, where similar dyes were also manufactured. This type of activity greatly improved the fortunes and performances of British dye firms. Such innovations were greatly assisted by technical education, with colour chemistry courses in Bradford, Glasgow, Huddersfield, Leeds, London and Manchester, which provided recruits for the 20th century dye industry, as did the courses of Raphael Meldola at the London Institution.

Indigo

After madder, indigo was the most important of the natural dyes. However, its synthesis proved to be extemely difficult. Unlike alizarin, it was not just a matter of substituting hydrogen atoms in a coal tar hydrocarbon. For indigo, it was necessary to construct the basic structural unit. By 1880, Baeyer had obtained sufficent information about the chemical constitution of indigo to devise a number of synthetic routes. Unfortunately, these were based mainly upon reactions of benzaldehyde, and the high cost of the starting material (toluene) was a major disadvantage from an industrial point of view. Both BASF and Hoechst supported the research of Baeyer, and were determined to be first in the race for indigo. Baeyer established the structure of indigo in 1883, and this enhanced the already considerable momentum that was leading towards commercialization of synthetic indigo. However, at the end of the decade Caro withdrew from the development effort, and Baeyer's activities in indigo research slowed down.

Interest now turned to Switzerland, where the growth of the dye industry had also been rapid. Rudolph Knietsch investigated synthetic indigo for CIBA in Basle, and later joined BASF to continue this work. Karl Heumann, at the Zurich Federal Polytechnic (ETH), also took up the challenge. In 1890 he discovered two methods, one based on aniline, and the other on naphthalene. Heumann's aniline method permitted the use of readily available benzene. The naphthalene process required conversion of this hydrocarbon to phthalic anhydride, and then into a derivative of phenylglycine. This second route held out the first hope of a commercially viable pathway to indigo, and, again, BASF and Hoechst shared the fruits of discovery.

Following seven years of expensive development - often claimed to have reached almost a million pounds sterling in cost - BASF produced synthetic indigo which could just compete with the natural substance. This relied on an efficient and economical method of oxidising naphthalene to phthalic anhydride with fuming sulphuric acid and mercury. Sulphur dioxide was evolved, and recovered, for conversion back into fuming acid by the contact process, that was also important to the indigo process. Knietsch brought the contact process to perfection, and BASF also acquired rights to the electrochemical production of chlorine and alkali, thereby ensuring that the most important inorganic chemicals were readily available.

BASF synthetic indigo became available to the dyeing trade in July 1897. Within three years, both BASF and Hoechst were also manufacturing indigo in France, where the product sold well because it was recognised to be easy to apply, and the uniformity of colour was maintained from batch to batch. By 1900, the quality and price of the German-made dyestuff had begun to reduce demand for

In the 19th century dyemakers proudly displayed their wares at international exhibitions. This is the BASF showcase for the 1885 exhibition in London. (Courtesy of Deutsches Museum, Munich).

the Indian-grown product quite noticeably. There was enormous concern in Britain because the major product of her largest colony, India, was threatened. Questions were asked in Parliament, and letters of concern appeared in the Times and in Nature. Central government intervened and demanded that all military wear be dyed with the product of Great Britain's "Jewel in the Crown." Despite the efforts to improve natural indigo production, many indigo users in Europe had gone over to the synthetic product by 1914. In Germany the military authorities stipulated the use of synthetic indigo when placing orders for clothing.

British - and American - Responses

The outbreak of the First World War revealed the almost complete dependence of the British textile industry on German dyes. Shortages were imminent, but the problem was overcome with with two remarkable success stories: James Morton's development of indanthrone vat dyes; and Herbert Levinstein's mastery of synthetic indigo.

Morton was a textile manufacturer and dye user based in Carlisle. In 1914, he employed a few chemists to convert anthraquinone-2-sulphonic acid, an intermediate in the alizarin synthesis, to yellow and blue vat dyes of the indanthrone type. This was achieved on a pilot plant scale, using high temperature and pressure in a steel autoclave. Dyes were produced in bulk one year later, followed by on-site manufacture of the intermediates. Morton built up a research and development team of the highest calibre, and in 1916 their efforts resulted in a fine substitute for the important alizarin blue wool dye. This was Morton's Solway blue. So remarkable were his products, made at first solely for use in his dyeing establishment, that they were soon in demand by other dye consumers, especially the Bradford Dyers Association.

Levinstein's achievement was to take over the requisitioned Hoechst factory at Ellesmere Port, and to produce phenylglycine, the essential intermediate for indigo, from aniline. Levinstein's synthetic indigo, the first made by a British company, became available by the end of 1916, and it soon satisfied the home demand.

Both Morton and Levinstein started almost from scratch in their efforts to emulate German discoveries, and their achievements were so highly regarded that the manufacturing methods were adopted in the United States, especially by the du Pont company. The pre-war American dye industry was hardly innovative, but the new conditions were a major spur to progress. For example, early in 1915, Herbert Dow began experiments on synthetic indigo, without consulting his fellow directors, and within eighteen months the Dow company was supplying its first customer.

The ability of British dyemakers to catch up with the Germans in certain vital products gave them enormous economic power in the post war world. In 1926 Morton's dye-making concern was absorbed by the British Dyestuffs Corporation (BDC), formed eight years earlier by a merger of the Holliday and Levinstein businesses. Later in 1926, BDC became part of Imperial Chemical Industries (ICI dyestuffs division, today ICI Specialties), created in response to the formation of

IG Farben in Germany. In 1931, ICI absorbed the British Alizarine Company, whose lineage could be traced back to William Perkin's pioneering factory.

Phthalocyanine, Fibre Reactive and Disperse Dyes

Chemical advances in the 1920s and 30s provided a deeper understanding of colour reactions, and suggested ways of improving both the design of colours, and their modes of adhesion to fibres. British products of the 1920s were modified direct dyes for viscose rayon, which were imitated in Switzerland and Germany. Several British chemists, including Arthur Green and William Perkin junior, sought out completely new types of dyes for British Alizarine, Levinstein, and Morton. The products included compounds that could be attached to novel fibres, like rayon and acetate.

Serendipity also remained a source of innovation. For example, chemists at ICI observed a blue colour in phthalimide prepared from phthalic anhydride and ammonia. It turned out to be the iron-containing analogue of natural pigments like chlorophyll and blood, and was soon transformed into Monastral fast blue (1934), containing copper in place of iron. This was the first phthalocyanine dye, and was followed up with other examples, especially a green.

In March 1956, exactly one-hundred years after William Perkin discovered mauve, ICI dyestuffs division placed on the market the first fibre reactive dye. These dyes undergo chemical reactions with the fibre to form covalent bonds and thereby induce unprecedented fastness. This was the beginning of the Procion range, ideally suited to cotton, and which gave rise to further research at CIBA in Switzerland, and at Hoechst in Germany.

A series of anthraquinone based colorants discovered at Blackley in 1923 was the first to be used for dyeing acetate rayon. These developments led directly to the most modern colours, the disperse dyes.

Disperse dyes are particularly important for the colouring of polyester fibres, which lack suitable chemical groups that permit bonding and admission into the fibres of dyes in aqueous solution. The problem is overcome by applying the dye as a fine dispersion, and the method has been adapted to other synthetic materials like nylon. Indeed, without the availability of this type of dye the manufacture of new synthetic fibres would never have been worthwhile. Disperse dyes became ICI's Suranol and Dispersol ranges. The latest disperse dyes are those based on benzdifuranone, and are ideally suited to use with polyester-cotton blends.

Finally, the fashion trend for denim since the 1960s has led to a massive resurgence in the demand for indigo, satisfied by major producers in the United States, Germany, Japan, and Britain. Modifications of the original processes are employed, but the raw material, as with most other synthetic dyes, is now prepared from oil.

New Uses for New Dyes

The development of synthetic dyes for clothing has inevitably led to an interest in other practical uses and in their chemistry. As early as 1862, cotton shortages in

Manchester caused by the American Civil War forced Roberts, Dale & Co. to seek out new outlets. Their mauve was used in letterpress printing and for colouring paper, and aniline colours were also used in postage stamps.

During the 1870s aniline and alizarin colours performed valuable roles in biological investigations as staining agents. In the 1880s, synthetic dyes were applied to examination of biological combustion. This led to ideas about cell structure and behaviour, and to how dyes and their analogues might be used to fight disease within the body. Extensive testing with azo dyes produced favourable results, but the side effects were unsatisfactory. The greatest success in this area was Paul Ehrlich's drug salvarsan of 1909, an arsenic equivalent of an azo dye. In the 1930s a red azo dye became the first of the sulphonamide range. By this time many former dye-making companies had turned their specialist skills to pharmaceuticals.

In more recent years the research into new coloured molecules has been stimulated by the electronics and communications industries with their needs for materials which absorb light, especially at the semi-conductor wavelengths. Novel derivatives of metal phthalocyanines have proved particularly useful. The rise in importance of office computers and copiers, as well as facsimile transmission machines, has led to the need for hard copy output on paper, especially colour graphics. This involves three dominant technologies: electrophotography (commonly called xerography), thermography, and ink jet. Whilst older dyes and pigments often suffice, the demand for high qualities requires new molecules specifically designed for the task.

Conclusion

Increased consumption of dyes, and new methods of preparation and application, led to many chemical based studies in the 1840s. In the late 1850s a new raw material, coal tar, became the basis of dyes made from aromatic amino compounds. This first generation of synthetic dyes was developed from mainly empirical studies. Increasing scientific knowledge and new theories in the 1860s and 70s enabled the synthesis of alizarin, and the introduction of the phthalein and azo dyes. Compounds with dyeing properties could also be predicted. By 1900, synthetic indigo had entered the market place, and soon brought about the disappearance of the natural dye. From the 1920s, British industrial chemists were at the forefront of developments that eventually led to fibre reactive and disperse dyes. One hundred and fifty years ago the fashion world relied on whatever colours chemists could wrestle out of natural products; today it is possible to provide tailor-made dyes to suit the demands of the fashion world.

The author wishes to thank Dr. Peter Morris, of the Open University, and Dr. Peter Bamfield, of ICI Specialties, for their assistance in the preparation of this paper.

Select Bibliography

J.J. Beer, The Emergence of the German Dye Industry. Urbana: University of Illinois Press, 1959; New York: Arno, 1981.

A. Bernthsen, "Heinrich Caro," Berichte der deutschen chemischen Gesellschaft 45, pt. 2 (1912): 1987-2042.

H. Caro, "Ueber die Entwickelung der Theerfarben-Industrie." Berichte der deutschen chemischen Gesellschaft 25, Referate, Patente, Nekrologe (1892): 955-1105.

M.R. Fox, Dye-makers of Great Britain, 1856-1976: A History of Chemists, Companies, Products and Changes Manchester: ICI, 1987.

W.M. Gardner, ed., The British Coal Tar Industry: Its Origin, Development, and Decline London: Williams & Norgate, 1915; New York: Arno, 1981.

L.F. Haber, The Chemical Industry During the Nineteenth Century: A Study of the Economic Aspect of Applied Chemistry in Europe and North America. Oxford: Clarendon, 1958, reprinted with corrections 1969.

A.S. Travis, "Science as Receptor of Technology. Paul Ehrlich and the Synthetic Dyestuffs Industry." Science in Context 3, pt.2 (1989): 383-408.

A.S. Travis, The Rainbow Makers: The Origins of the Synthetic Dyestuffs Industry in Western Europe. Bethlehem, Pennsylvania: Lehigh University Press, 1992.

Materials

Chemistry and the Production of Metals

J.K. Almond

TEESSIDE POLYTECHNIC, MIDDLESBROUGH TS1 3BA, UK

1 INTRODUCTION

When, in 1841, the Chemical Society of London was founded, the nearby government-supported Museum of Economic Geology was advertising for its first pupils to receive 'instruction in analytical chemistry, metallurgy, and mineralogy' from the curator and chemist Richard Phillips, FRS. Part of his job was to analyse ores and soils for the public, at moderate charges.[1] Ten years later, in 1851, the teaching function of the museum was formalised by the creation of the Government School of Mines, later known as the Royal School of Mines. The lecturer on metallurgy at the school was the redoubtable Dr John Percy, who had qualified in medicine at Birmingham. For the next thirty years the School of Mines in London was the only place in England and Wales which offered worthwhile instruction in the field of 'metallurgy', although the subject was then largely confined to metal production, with little stated about either the arts of combining metals to give alloys possessing enhanced properties or the crafts of forming metallic materials into desired shapes. As the result of fortunate powerful interventions at several times of crisis, the Royal School of Mines continues today as a constituent part of Imperial College. The courses have changed somewhat since 1851, and especially is this true of the period since 1951. Meanwhile, during the last hundred years instruction in aspects of the 'metallic arts' became available at many other academic centres; and, since the boom years of the 1950s when the use of atomic energy was attracting public funds, a fair number of such courses have died.

Table 1 Quantities of metals produced world-wide

10^6 tonnes a year

Year	Non-Ferrous Metals	Iron and Steel	Total
1841	0.16	3	3.16
1915	4.15	65	69.1
1989	45 plus	780	825 plus

As for the metals themselves, only in rare instances do metallic elements occur naturally in uncombined form, as 'native metal', some gold being a good example. In all other circumstances, in order to yield useful metals from raw resources, chemical reactions are necessarily involved. In fact, without the application of chemistry, there would be no significant world metals industry. Currently this industry is large, with metals (as a whole) being produced to the extent of at least 825 million tonnes a year, an average of 150 kg for every human being. This enormous present tonnage contrasts with the 3 million tonnes which was the corresponding figure in 1841, Table 1. The year 1915 represents the half-way point in the 150-year period. The tonnage figures show growth factors, over the 150 years, of more than 200/1 for both non-ferrous and ferrous sectors of the industry.

The number of metals produced in substantial quantities has also increased during the 150 years, from about 9 to more than 16, Table 2.

Table 2 Metals produced industrially

already produced in 1841:	iron copper lead tin mercury silver gold platinum zinc
introduced on large scale since 1841:	aluminium magnesium sodium nickel tungsten titanium uranium

Chemistry is only one of several factors which are involved in the successful production of metals from their raw mineral resources, and by itself the chemistry, no matter how favourable it is, is ineffectual; it needs to be channelled and guided so that the desired result may be obtained. Generally, it is necessary that the chemical reactions be followed by appropriate phase separations. As an example, consider the typical reaction situation in which lead oxide is reduced to metallic lead, a process carried out at a temperature sufficiently high to give fluid metal and slag:

$PbO_{(s)}$	+	$SiO_{2(s)}$	+	$CO_{(g)}$	+	$CaO_{(s)}$
oxidised ore mineral		impurity accompanying ore mineral		from partial combustion of carbon		flux deliberately added

=	$Pb_{(l)}$	+	$CaO.SiO_{2(l)}$	+	$CO_{2(g)}$
	wanted metal		slag		gas

The products can be removed from the reactor because they are fluid, and the wanted metal can be separated from the other phases present. (However, the wanted metal is likely to contain some impurities and so to require a subsequent refining stage.)

2 TWELVE APPLICATIONS OF CHEMISTRY TO METAL PRODUCTION SINCE 1841

The importance of chemistry for metal production during the 150-year period since 1841 can be illustrated by the following twelve major applications, which are outlined more or less in chronological order of successful industrial production resulting from their introduction.

Removal of some impurities from crude molten iron by blowing air through the melt at 1200-1300°C, with generation from oxidation reactions of sufficient heat to raise the temperature to more than 1500°C. This was Henry Bessemer's great contribution, worked out between 1855 and 1860. It is considered in more detail later in the chapter.

Removal of silver from molten lead by addition of zinc, forming silver-zinc compound immiscible in lead. This method was introduced by Alexander Parkes of Birmingham, who secured a patent as early as 1850. However, it was only after 1860 that the scheme was favourably received, and began to be taken up increasingly by lead refiners. Among the obstacles to successful application was the problem posed by the excess of zinc, which remained in the lead as an unwanted impurity; eventually at least four alternative ways came into use to solve that difficulty.

Removal of phosphorus from molten iron by use of lime. The major shortcoming of Bessemer's process was that it would not deal with phosphorus. In the period 1878-1880 a solution was presented by Sidney Thomas (1850-1885) and Percy Gilchrist (1851-1935), who developed for the reactor a refractory ceramic lining based on hard-burned (ie calcined) dolomite; this was compatible with the abundant lime which they added to the melt and which 'fixed' the phosphorus as calcium phosphate after its oxidation, so preventing its reversion from slag to metal. The 'basic' method of working supplied the key that enabled crude irons containing 2 percent of phosphorus to be treated to yield marketable engineering steels.[2]

Production of liquid aluminium by electrolysis of alumina dissolved in fused fluorides of aluminium and sodium. The electronic-reduction process was developed during the years 1885 to 1889 by two men, working totally independently of one another. Charles Martin Hall lived in the USA, while Paul Héroult was based in northern France. Both were in their early twenties when they carried out the inspired experimental work which led them to their separate patents. Indeed, both men were born in the same year, 1863, and both died in 1914. Hall and Héroult were able to offer aluminium at a small fraction of its former price, and from about 1890 (only five years after their experimental work had begun) the electrolytic method started to be adopted in earnest, and aluminium, from being a minor metal and curiosity, became a major world player.

Economic production of alumina trihydrate from bauxite by use of (a) hot aqueous sodium hydroxide to take aluminium into solution as aluminate, and (b) seed crystals to deposit from solution coarsely-crystalline alumina trihydrate. These two joint applications of chemistry are still firmly associated with the name of Karl Josef Bayer, an Austrian chemical engineer who introduced them just a year or two after the Hall-Héroult reduction method appeared. Compared with earlier ways of treating bauxite to extract its content of aluminium, Bayer's scheme offered significantly higher extraction efficiency, coupled with a filterable solid product.

Concentration of gold from ores by dissolution in aqueous alkali cyanide followed by its precipitation by contact with zinc. Some of the component strands for the chemistry of 'the cyanide process' for extracting gold date from the second and third quarters of the 19th century, but it was not until the years 1885-1890 that the method was developed into something that became commercially successful. The mainspring behind this chemical application was John Stewart MacArthur (1856-1920), a Glasgow chemist who was employed by Tennants and was therefore familiar with the large-scale 'wet chemistry' practised to recover copper, and sometimes silver, from cupriferous pyrites.[5] From the 1890s the method became increasingly important in world production of the precious metal.

Separation of nickel from copper by use of carbonyl compounds, nickel forming with CO a gaseous tetracarbonyl while copper does not. It seems to have been in 1889 that Carl Langer (1860-1935), working in Ludwig Mond's laboratory in London, observed the decomposition of nickel carbonyl with the deposition of metallic nickel. Mond built an experimental, demonstration, plant in 1892 but it took another ten years before the method was taken up industrially - and then by Mond himself, who purchased Canadian mineral deposits in order to obtain a feed for his separating plant, built at Clydach in South Wales. Until then, because of their generally similar properties, it had proved difficult to separate copper from nickel arising from the extensive deposits in which both elements occurred as sulphides, intimately associated.

Production of zinc by (a) leaching oxidised minerals with aqueous sulphuric acid, followed by (b) electrolytic deposition of zinc from sulphate solution. Contributions towards a successful application were made over the period 1880-1914 by at least six people living in various countries. The commercial process was introduced in North America in 1914, at a time when the onset of war in Europe led to increased demand for high-purity zinc for armaments; thus there was financial stimulus.

Figure 1 Henry Bessemer in later life; reproduced from a portrait photograph given by the inventor to Richard Smith, the metallurgical assistant at the Royal School of Mines who carried out practical analyses. (courtesy of Dr A.E.W. Smith)

Reaction of calcined dolomite with silicon at 1200-1400°C, under vacuum, to yield gaseous magnesium. Small quantities of magnesium had been produced during the second half of the 19th century by the reduction of magnesium chloride with sodium, and by the 1930s an electrolytic process, similarly based upon anhydrous magnesium chloride, was also producing metal. However, the electrolytic route was expensive, and it involved corrosive gaseous chlorine. It was against this background that German workers, led by the energetic Dr Gustav Pistor, in the 1930s succeeded in extracting magnesium from the abundant calcined dolomite by reaction with silicon (The relevant experimental work can be dated from ca.1915).[4] In practice, to obtain an industrial yield of metal, there were some formidable barriers to be overcome; the process was carried out at much reduced pressure (ie in vacuum), while the presence of lime in the doloma was advantageous because of its great affinity for silica. In North America the method was introduced in 1941 by Dr Lloyd M. Pidgeon. It has become a text-book classic on account of the thermodynamics and kinetics involved.

Reaction of titanium tetrachloride with either sodium, calcium or magnesium to yield titanium. (Similar procedure is used to obtain zirconium and uranium.) The chemical roots of this application were well documented in the 1880s, but it was only during the 1940s that industrial production began on any significant scale. Along the way, M.A. Hunter, working for the General Electric Company in the USA, in 1910 produced titanium by sodium reduction. During the 1930s in Europe, and then in the 1940s in the USA, Dr William J. Kroll developed metal production using first the expensive calcium and second the cheaper magnesium. Regular production of titanium metal can be considered to have started in 1948.

Presentation of some thermodynamic information in comprehensible form, namely charts of standard free energy of formation plotted against temperature, ie ΔG^{o} formation versus T. This application, introduced during the 1940s, is considered more fully below.

Use of anion-exchange resins to concentrate and purify uranium compounds from large volumes of weak solutions resulting from leaching. World demand for uranium was keen during the 1940s, and government funds were poured into development work on treatment methods. It was then already known that most metals, when in aqueous solution, would be absorbed by cation-exchange resins. It appears to have been largely by accident, however, that in 1948 workers at two places in the USA, the Battelle Memorial Institute and the Dow Chemical Company, discovered that, almost alone among metals (ferric iron being another important exception), uranium present in sulphate and carbonate solutions could be deposited onto anion-exchange resins.[5] Soon it was realised that this finding provided a technique of great selectivity and considerable value. In 1950 an international team (representing both USA and British interests) introduced the new technique on pilot-plant scale at a South African uranium producer; in 1952 the first large-scale plant was commissioned, and in the following years substantial proportions of world production have been treated.

Figure 2 Bessemer converters for refining crude iron to steel, introduced c1859. ('Sir H. Bessemer FRS - an autobiography ...', Engineering, London, 1905, plate XV, facing 148)

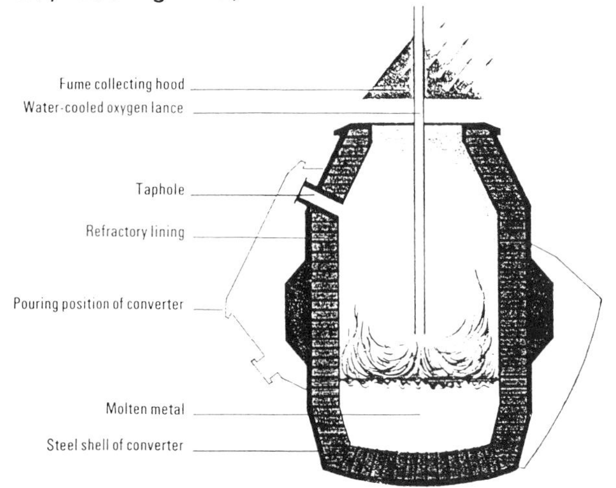

Figure 3 L-D top-blown oxygen converter c1970. (Reproduced with permission of British Steel Plc)

Comments on the Foregoing Chemical Applications

Eight of the twelve applications cited belong to the 75-year period between 1841 and 1915, while all of the remaining four had taken place by 1955. It seems clear that the most active development period in this particular field was the second half of the 19th century.

For any application of science to industry to succeed there must be a convergence of four things: (1) an appropriate idea; (2) suitable technology, to enable the idea to be carried out industrially; (3) incentive to human endeavour; and (4) a favourable economic climate. Sometimes it happens that all four conditions come together at much the same time, when a rapid positive result ensues. Of the applications given above, several reached a successful stage quickly, that is within four to five years of the start of experimental work. Examples are Bessemer's converting process for steel, Hall and Héroult's production of aluminium by electrolytis, and the use of anion-exchange resins to concentrate uranium.

On the other hand, in general there appear to be more instances in which some of the conditions necessary for success become available many years before the remaining prerequisites are fulfilled. Thus, with several of the foregoing applications of chemistry to metal production there was a time interval of more than 25 years before all the conditions converged. In the extraction of zinc by hydrometallurgy using aqueous sulphate followed by electrolytic reduction, limited aspects of the system were developed over a period of 34 years before the whole thing was brought together and, even then, the industrial process that was introduced was to undergo many subsequent modifications. A second application with a long lead time is the production of titanium by reduction of the tetrahalide with sodium, calcium or magnesium. The chemical reaction was tried in 1887, but it was only applied on a significant industrial scale after 1940, that is half a century later.

Figure 4 H.J.T. Ellingham, FRIC, in 1939. (Royal Society of Chemistry)

3 HENRY BESSEMER AND 'CONVERTING' PROCESSES

It was only fifteen years after the launching of the Chemical Society that Henry Bessemer created a spectacular milestone in the chemistry of metal production. This related to his 'converting' process whereby molten impure iron from a blastfurnace could be transformed into the purer metal subsequently known as 'bulk steel', which possessed superior engineering properties. The impure feed was liquid when charged to the equipment at a temperature of 1200-1300 $^{\circ}$C, while the resultant product was also liquid and at the higher temperature of more than 1500 $^{\circ}$C. The refining was done without the use of external fuel; all necessary heat for the process was obtained from exothermic reactions between atmospheric oxygen and components present in the metal bath, notably silicon and carbon.

Henry Bessemer (1813-1898), born at Charlton in Hertfordshire, by the early 1850s was already a successful inventor with a factory in London making profitable imitation-gold powders for bookbinding and other decorative purposes, Figure 1. Because of the Crimean War he devised a new spinning projectile that interested the French authorities. However, the projectile demanded better material than castiron for the barrel of the gun from which it was to be fired, so, at the beginning of 1855, Bessemer began experiments in his factory with the aim of producing superior metal. A chance observation led him to think that the crude blast-furnace iron he was using was being decarburised by air jets which he had fitted into the furnace to improve the fuel combustion. Small-scale tests confirmed this supposition but also, to his surprise, revealed that as air was blown through the metal bath the temperature increased considerably. Having perceived what was happening, Bessemer possessed both the ability and the resources to try to profit from the observations, and he continued with experiments.

Bessemer presented an account of his new process, and all that it promised, to the 1856 meeting of the British Association for the Advancement of Science, held in Cheltenham, and ensured that The Times carried a full report. He entitled his paper 'On the manufacture of iron without fuel'. As he had hoped, several industrialists came forward seeking permission to use the new process, and some hard business deals were struck over licences, Bessemer and his partners apparently receiving £27 000 from potential users within the space of four weeks. However, when the licensees tried to introduce the new technique in their own works, all without exception found it impossible to make good metal. Soon the tide of opinion turned against the inventor and his process. The British Association decided to omit his Cheltenham paper from the volume of Transactions.

With the benefit of hindsight, we know that Bessemer had been extremely lucky in the kind of impure iron which he had innocently purchased for his experiments; it contained the right proportions of silicon, carbon and manganese to behave in a favourable way, while it carried only small proportions of the two troublesome elements phosphorus and sulphur. Other people were not so fortunate with their irons!

He soon learnt to use only those crude irons which had a low (ideally less than 0.1 percent) phosphorus content. Essential advice concerning the effective way of introducing the cold airblast into the molten metal was given to him by a Swedish licensee G.F. Göransson, who had discovered it during desperate experimentation. The valuable benefits to be gained by adding manganese and carbon at the end of the 'blow' were drawn to the inventor's attention by a rival, Robert Mushet. With business partners, Bessemer erected works at Sheffield to produce high-carbon steels in competition with the established steelmakers, and to serve as a demonstration unit. For these works, c1859 a new pattern of reaction vessel was introduced in place of the fixed 'converters' used in London and elsewhere, Figure 2.

From about 1860, interest was renewed in what the inventor had to offer.[6] Equipment to use his patented process was erected in the chief industrialised countries, and in each successive year increased quantities of the new engineering material were produced. In 1869/70 world output of bulk steel was around 0.4 million tonnes, while by 1875 it exceeded one million tonnes. Bessemer's technique led the way in the industrial progression from 'iron' to 'steel': in place of the iron which had already made possible the firm beginnings of industrialisation, with steam engines, machinery and railways, he offered bulk steel as a better material.

In the second half of the present century world steelmakers have increasingly adopted the injection of oxygen in place of air, despite the fresh problems initially created by the substitution. With neat oxygen blown into the melt from below, the temperatures generated are locally so high that rapid attack occurs at the ceramic bottom of the reaction vessel. Two different ways of overcoming this difficulty have been developed. In bottom-blown oxygen converting (eg the German OBM process) the gas is injected into the molten-metal bath as in conventional Bessemer working, except that here the oxygen streams are sheathed in gaseous hydrocarbon which exerts a cooling effect. In top-blown converting (ie the L-D and BOS processes), the oxygen is jetted downwards into the molten bath from a lance tip held 1-2 metres above the surface, Figure 3. L-D converting is the predominant method for making bulk steel from molten impure iron, but a modification which is gaining ground is the addition of small gas-injection ports at various positions in the reaction vessel to provide better mixing of the bath and improved heat generation.

A large present-day L-D converter can deal with a batch of 300 tonnes of metal in about 45 minutes, of which some 12-18 minutes are occupied by the actual 'blow'. Thirty percent of cold scrap steel can be assimilated into the melt but, even so, the temperature of the product, poured from the vessel when it is tilted, is close to 1600°C.

During the 1880s Bessemer's 'converting' method began to be applied to the production of copper from mixed copper-iron sulphides. Here the heat generated by the selective oxidation of iron sulphide is usefully employed to maintain fluidity in the contents of the reaction vessel and to form a slag phase containing the oxidised iron, so that separation may be made. In recent years the principle has been extended further, to assist in the extraction and refining of several other metals.

4 ELLINGHAM, RICHARDSON, AND 'FREE-ENERGY DIAGRAMS'

In marked contrast with the other applications of chemistry, which are to do with the treatment of particular metals for specific commercial results, the exposition, made during the 1940s, of certain thermodynamic information in diagrammatic form is a notable presentation of theory: it provides a powerful tool for predicting the conditions that are necessary for particular reactions to be thermodynamically favourable. With this clear setting out of information two people are most closely identified, Dr Ellingham and Dr (later Professor) Richardson. Apart from the intrinsic value of their work in this field, it is appropriate to mention these two men at an event organised by the Royal Society of Chemistry, and also in the context of Imperial College.

Bessemer's 'converting' process depended totally upon thermo-chemistry, and recognition of the importance of the energy involved in unit operations grew during the second half of the 19th century and the first half of the present one. By the 1930s, investigators in Germany and the USA were busily collecting experimental thermochemical information and attempting to digest it into forms that could have relevance for industrial activities such as steelmaking and zinc production. It was Dr H.J.T. Ellingham, an electrochemist and assistant professor in the chemistry department of the Royal College of Science, a constituent part of Imperial College, who showed in an elegant and easily-

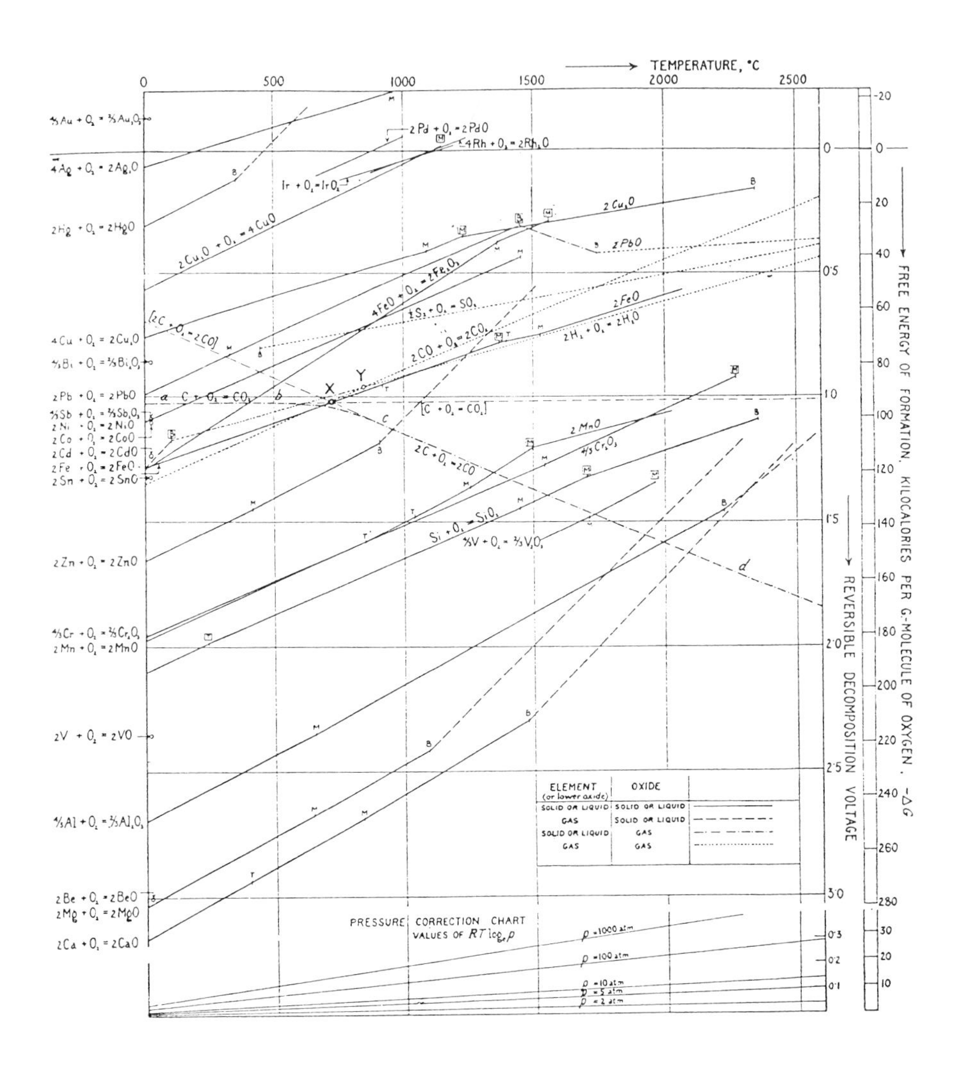

Figure 5 Ellingham diagram for metal oxides, 1944. (Reproduced, with permission, from J. Soc. of Chem. Ind., 1944, 127)

Figure 6 F.D. Richardson, FRS. (Reproduced from J. Iron Steel Inst., 1968, 206, with permission of the Inst. of Metals)

comprehensible diagrammatic way the relative standard free-energy changes involved in a family of similar reactions.

H.J.T. Ellingham (1897-1975), Figure 4, was born in London and graduated at the Royal College of Science in 1917 with a first-class honours degree in chemistry, after which he was commissioned in the Royal Engineers and sent to Mesopotamia (ie, part of present-day Iraq) as a water chemist. By 1919, however, he was back in the chemistry department in South Kensington as a demonstrator, from which level he gradually progressed through the academic heirarchy; in 1937 he was appointed reader in physical chemistry in the University of London. Between 1934 and 1939 he served as honorary secretary of the Chemical Society.[7]

Part of Ellingham's work was to teach physical chemistry to students of chemical engineering and metallurgy. Like other good teachers, he put effort into trying to produce order out of chaos, to categorising populations of facts. He was a master of the chart as a means of showing the relationships between individual components and for communicating information. In this case, for a number of metals he considered their reactions with oxygen, and calculated for standard conditions the associated free-energy change occurring over a range of temperature. These results he presented graphically.

Figure 5 is reproduced from Ellingham's article published in 1944, in which he presented diagrams for two families, metal oxides and metal sulphides. A footnote to the article states that its contents were based upon a lecture which Dr Ellingham delivered at Swansea in 1943, 'at a joint meeting of the Chemical Society, the Royal Institute of Chemistry, and the University of Swansea Chemical Society'. For some time before then Ellingham must have given thought to an effective way of presenting the information.

In 1948 Ellingham had another opportunity to publicise his scheme when the Faraday Society organised a discussion on 'The physical chemistry of process metallurgy'; he was the joint author of papers presented at the meeting, and he participated in the discussions. However, by this date, after a quarter of a century in South Kensington, Ellingham had left his secure academic position to become the secretary of the Royal Institute of Chemistry, based in Russell Square. In 1952 he took on the additional post of registrar, remaining secretary and registrar of the Institute until his retirement ten years later. In 1973 the comment was made that Dr Ellingham 'fanned a few sparks into flames when he introduced his free energy diagrams and made free energy and entropy mean something to process metallurgists.'[(8)]

Among those who soon recognised the value of Ellingham's method of presentation was Dr F.D.Richardson, then working in London for the British Iron and Steel Research Association (BISRA), Figure 6. Denys Richardson (1913-1983), was born in Streatham on the south side of the capital, graduated in chemistry at University College London, and obtained his PhD there in 1936 for work on the unstable chlorine oxides. Throughout the war of 1939-45 he served in the Admiralty's Department of Miscellaneous Weapons Development, becoming deputy director with the rank of commander. At the end of the war he joined the new

BISRA as superintending chemist. At first there were no laboratories in which Richardson could initiate his own experimental work but, following a meeting with Dr Ellingham and helped by his BISRA staff, he set about the preparation of several new free-energy diagrams together with refinement of Ellingham's diagrams for oxides and sulphides, incorporating the best thermodynamic data available. In 1949, Richardson and J.H.E. Jeffes introduced a new feature on the charts, modifying Ellingham's limited pressure-correction scale into a nomograph covering a wide range of pressure deviations from one atmosphere. Figure 7, originally published in 1952, shows some metal-sulphur combinations. The nomographic scale enables both the equilibrium pressure of sulphur (pS_2) and the equilibrium ratio of H_2S/H_2 to be read off at any temperature for any of the stated reactions.

By 1951, other workers had published 'Ellingham diagrams' for systems of metals and their compounds that included chlorides, fluorides, carbonates and silicates.

Denys Richardson in 1950 transferred his energies from BISRA to a new Nuffield fellowship set up with £5000 to encourage investigation into the fundamental physico-chemical phenomena involved in extracting metals. He was already interested in slags, and the constitution and behaviour of these substances, together with their interactions with metals, became important topics for study and elucidation. The Nuffield Research Group in Extraction Metallurgy, headed by Richardson, was housed in the Royal School of Mines, in South Kensington. In 1957 Richardson was appointed professor of extraction metallurgy at the school. When, 19 years later, he retired, he had a well-established international reputation: he was a leading exponent of the 'new look' in the chemistry of metal extraction. For a quarter of a century his research groups attracted workers from many parts of the world. In 1968 he was elected a Fellow of the Royal Society.[9]

Both Ellingham's original concept, and the large amount of information contained in Richardson's elaboration, were quickly and widely grasped, not least by educators and text-book writers. From 1950, by use of this tool, books on metal extraction were given new academic respectability: the old descriptive style of text was entirely replaced by one which dealt with

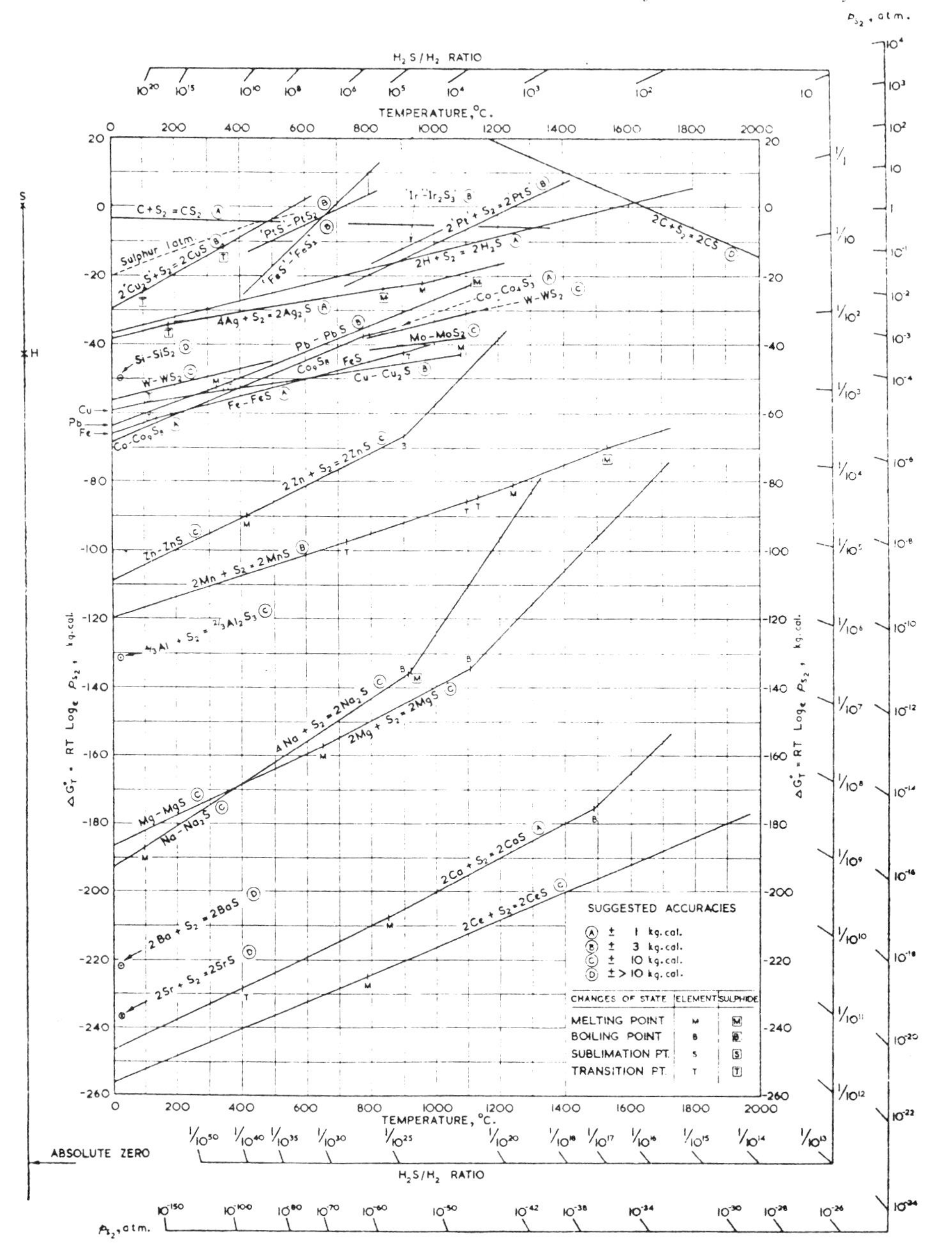

Figure 7 Ellingham-Richardson diagram for metal sulphides, 1952. (Reproduced from J. Iron Steel Inst., 1952, 171, 167, with permission of the Inst. of Metals)

principles, whereby the theoretician was given an advantage over the practical craftsman. More positively, throughout the last forty years Ellingham-Richardson diagrams have led to far wider understanding of the thermodynamic principles underlying the reactions involved in metal production, while they have enabled predictions to be made, quickly and cheaply, concerning the likely feasibility of any proposed reaction. The diagrams constitute a significant forward step in the chemistry of metal production.

5 CLOSING SUMMARY

I have listed eleven different ways which have been introduced in the years since 1841 in order to yield specific metals, or refine them, by means of chemical reactions. All of these chemical applications remain important today, in most cases accounting for large proportions of the total outputs of the metals with which they are concerned.

Thus, Bessemer's innovative application of thermochemistry, in its later oxygen-based modifications, is currently responsible for about 400 million tonnes of world steel each year, more than one half of all the steel that is produced. Of this total, probably 700 million tonnes, nine tenths, is made in the 'basic' conditions introduced by Thomas and Gilchrist. As far as copper is concerned, some 7½ million tonnes a year, four fifths of the total primary metal made, result from treatment that includes a 'converting' stage. Turning to aluminium, virtually all of the 18 million tonnes of primary metal is derived directly from the processes brought to fruition by Hall, Héroult and Bayer. With gold, the major fraction or world output comes from the 'cyanide process' of MacArthur; taking a gold price of £200 a troy ounce, the value of the gold so produced amounts to more than £6000 million a year. Of world zinc, about two thirds, or nearly 5 million tonnes, is the result of leaching and electrolysis. Lastly, to move to the example of a metal which only began to be produced in significant quantities close to 1950, the world titanium output of the order of 100 000 tonnes a year is all obtained by use of the chemistry developed by Hunter and Kroll.

Altogether there is abundant evidence that, so far, chemistry has played a vital part in the production of world metals. By doing so, chemistry has contributed substantially to the improvement in living standards which has come about by the application of metals to so many of man's activities. There is every reason to think that, in the future, this key funtion of chemistry in metal production will continue unchanged.

SOURCES CITED

1. Sir T.C. Chambers, 'Register of the associates and old students ... of the Royal School of Mines ...' Hazel, Watson and Viney, London, 1896, ix; quoting Buckland, Proc.Geol.Soc., 1841, 3, 211.
2. J.K. Almond, Ironmaking steelmaking, 1981, 8, 1-10.
3. D.I. Harvie, Endeavour, new series, 1989, 13, 179-184.
4. F. Fox and G. Lewis, Jrnl.Histor.Metallurgy Soc., 1985, 19, 63; A. Beck, 'The technology of magnesium and its alloys', transl. L.H. Tripp and others, F.A. Hughes, London, 1940, 16-17.
5. T.V. Arden, in 'Extraction and refining of the rarer metals', Inst. Mining Metallurgy, London, 1957, 119-129.
6. K.C. Barraclough, 'Steelmaking: 1850-1900', Inst. of Metals, London, 1990.
7. F.W.G. Inst.Chemistry Procs., 1963, 87, 31-34; G.J.H. and L.W.W., Chem.in Britain, 1976, 12, 322-323.
8. Sir C. Goodeve, in 'Physical chemistry of process metallurgy: the Richardson conference ... 1973', eds. J.H.E. Jeffes and R.J. Tait, Instn.Mining Metallurgy, London, 1974, xi-xii.
9. J.H.E. Jeffes, Biogr.mems. fellows Royal Society, 1985, 31, 495-521.

Some Milestones in Plastics

Thomas R. Manley

2 MOORFIELD, NEWCASTLE UPON TYPE, NE2 3NL, UK

Natural rubber (poly _cis_-isoprene) was known in Spain in 1550 and a century later Tradescant introduced gutta percha (poly _trans_-isoprene) to Britain. Hancock between 1820 and 1843 masticated crude rubber to make it plastic and then "vulcanized" it with sulphur to give an elastic material. If more (30%) sulphur was used "hard rubber" or Ebonite was obtained. This was the first thermosetting plastic. The boundaries between rubbers and plastics are ill defined: thermoplastic elastomers, polyurethanes, silicones even highly cross linked polythene may be elastomeric or plastic. Rubber is worthy of a chapter on its own so we will restrict ourselves to plastics.

In 1861 Parkes patented "Parkesine" and in 1862 at the International Exhibition held in Hyde Park, London, Parkes was awarded a medal for his new material. The following quotation from the official catalogue Class IV No 1112 shows why 1862 is taken as the start of the plastics industry (1).

PARKESINE

"A new material now exhibited for the first time has from its valuable properties induced the inventor to patent the discovery in England and France and to devote the last ten years to the development of the application of this beautiful substance to the Arts. In the Case are shown illustrations of the numerous purposes for which it may be applied such as MEDALLIONS, SALVERS, HOLLOW WARE, TUBES, BUTTONS, COMBS, KNIFE HANDLES, PIERCED AND FRETWORK, INLAID WORK, BOOK BINDING, CARD CASES, BOXES, PENS, PEN HOLDERS ETC. It can be made HARD AS IVORY, TRANSPARENT OR OPAQUE, of any degree of FLEXIBILITY and is also WATERPROOF, may be of BRILLIANT COLOURS, can be used in the SOLID, PLASTIC or FLUID STATE, may be worked in DIES and PRESSURE as METALS; may be CAST or used as a COATING to various substances; can be spread or worked as INDIA RUBBER and has stood exposure to the atmosphere without change or decomposition. By the system of ornamentation patented by Henry Parkes in 1861 an endless variety of effects and the perfect imitation of TORTOISE SHELL and WOODS may be produced."

The key discovery that Parkes made was that camphor has a plasticising action on cellulose nitrate. This was probably Parkes greatest contribution to technology even though his process for the removal of silver from lead was used for over a century and he patented in 1846 the process of cold vulcanization of rubber. Possibly because of these other interests the work on Parkesine was not pursued.

However, around this time the price of African ivory was rising steeply because of the increase in hunting elephants and a New York firm Phelan & Callender offered a prize of 10,000 dollars for an acceptable substitute for ivory.

In 1869 in America J.W. Hyatt took out a patent on "Celluloid" which was a solution in ethanol of cellulose nitrate and camphor. Celluloid was a commercial success and remained in production for a century although Hyatt's patent was found to be invalid because of Parkes' earlier work. A year after the Great Exhibition on 14th November 1863 Leo Hendrik Baekeland was born in Ghent. He was a brilliant student and a keen amateur photographer, his first patent in 1887 was in the field of photography. Baekeland obtained a travelling scholarship and was keen to do research in Britain but he had to settle for the USA to the great benefit of that country and loss to Europe.

Initially he continued his photographic work inventing "Velox" gaslight photographic printing. His company was bought out by Eastman Kodak leaving Baekeland financially independent but pledged to keep out of the photographic business.

In 1872 Bayer had obtained a resinous product from the condensation of phenol with acetaldehyde. Classical organic chemists were mainly interested in crystallizable materials so these phenolic resins were usually regarded as intractable and discarded.

In 1899 however, A. Smith took out the first patent on phenol aldehyde resins (2) for use in electrical insulation.

Baekeland used acid catalysts and a low ratio of formaldehyde to phenol to make "Novolacs" (i.e. new shellacs) which were permanently fusible and soluble in acetone or alcohol. These replaced Shellac as electrical insulation and varnishes. Using basic catalysts Baekeland obtained resins in three stages A, B and C.

"A" stage resins resembled Novolacs but on further heating (i.e. more cross linking) gave "B" stage resins, gelatinous when hot and solid when cold. Further heating converted B stage resins to the hard, insoluble, infusible "C" stage. This was reported in Baekeland's historic patent of 18th February 1907.

The first castings were spongy and porous (water of condensation is a by product of the P/F reaction) so in the next patent (July 1907) Baekeland poured the "A" resin into a mould and cured it directly to the "C" stage under external pressure at 150°C; thus moulding could be performed in minutes instead of the hours required at 100°C. The clear P/F resins found uses in artificial jewellery but were too brittle for other uses. However, P/F resins readily combine with sawdust to give products of moderate mechanical strength and by the use of other fillers, e.g. mica, cotton, nylon, silica and minerals, resins with a wide range of properties may be made.

"Bakelized paper" was extensively used as electrical insulation (e.g. bushings) before the advent of epoxy and is the basis of Tufnol (gears & engineering plastics) and printed circuits. Other phenols and aldehydes are used as wood adhesives. The Mosquito bomber was held together with resorcinol formaldehyde glues. Baekeland announced his results in February 1908 in New York and considerable interest in the new materials was generated.
There had been however some interest in P/F resins in Britain. Sir James Swinburne, born in Inverness on 28th February 1858, trained on Tyneside as an electrical engineer. There he met Joseph Swan the inventor of the electric light, which was first demonstrated in the Lit & Phil in Newcastle at a meeting of the Newcastle Chemical Society. Swan established a factory in Paris in 1881 to make carbon filament electric lamps with Swinburne in charge.

In 1902 Swinburne became President of the Institute of Electrical Engineers and in his presidential address he drew attention to the need for better insulation for cables. In search for better insulation he reacted controlled quantities of phenol and formaldehyde and obtained useful products. He formed the Fireproof Celluloid Company in 1904 to exploit this material but carried on work to improve the resin and delayed applying for a patent until 1907. In the end Swinburne found that Baekelite had anticipated him by a single day.

Swinburne first turned to the hard lacquers used to protect brass forming the "Damard" Lacquer Co which described the main property of the resin, but in 1916 he decided to licence Baekeland's patents and in 1928 Bakelite was formed with Swinburne as Chairman to amalgamate all the British companies in the P/F field.

Between the wars Bakelite was the main plastic in Britain and elsewhere. The electrical industry was transformed by bakelized paper and bakelite electrical fittings replaced brass. Engineering plastics and records were made from Bakelite.

At this period as a by product of the use of cellulose acetate dope for aircraft the long search for a non-inflammable celluloid ended. Cellulose acetate sheet became the first plastic packaging material and the first injection moulded material. CA is hard, tough and readily moulded so it still is used, e.g. in electrical/screwdriver handles and as a packaging or photographic film. Between the wars CA was the only thermoplastic; most of its applications have now been taken by PS (Polystyrene), PE (Polythene) and PP (polypropylene). In Germany PS was the main thermoplastic.

At the beginning of the century the milk protein casein was reacted with formaldehyde to give white products that could readily be coloured. At one time "Galalith" and "Erinoid" had many applications but nowadays buttons are practically the only commercial application. Between 1920 and 1930 first extruders and then injection moulding machines were developed. These machines enabled rapid production of moulded articles from thermoplastics and led the way to mass markets. Polymers are usually dearer than traditional materials but the ease of processing usually means that the finished object is cheaper when plastics replace older materials.

Other Condensation Polymers

The success of P/F resins fostered interest in other resins, using urea, melamine, aniline and other phenols and other aldehydes such as furfural.

Urea formaldehyde resins being clear and transparent were thought to be possible organic replacements for glass but they are too susceptible to moisture. U/F resins have largely replaced P/F where appearance is the main characteristic - pastel shades of electrical fittings replaced the black P/F mouldings. U/F resins are widely used as adhesives e.g. in the manufacture of chipboard. U/F foams provide thermal insulation in houses built with cavity walls.

In May 1922 Staudinger showed that the concept of polymers as some form of colloidal dispersion was erroneous. He introduced the term "macromolekul" to describe polymers, and "polystyrene" instead of "metastyrene". There was considerable opposition to Staudinger's ideas but he overcame the opposition by further research and in 1953 he was awarded a Nobel prize.

A decade after the publication of Staudinger's historic Paper I.G. Farben in Germany and Dow in the USA began the commercial production of polystyrene. Initially PS was used as an electrical insulator, e.g. in capacitors, but the ease of moulding of PS and the transparency of the resin have led to PS becoming the fourth largest plastic in terms of tonnage used. Politics have played a large part in the history of PS. Hitler encouraged its production before the second world war; the loss of Malayan rubber in 1942 led to the production of synthetic rubber in the USA. Ironically the process used was that developed by I.G. Farbenindustrie to make Buna (Butadiene-sodium) rubber. Government rubber-styrene or GR-S was made in vast quantities during the war but became obsolete after VJ Day and a market had to be found for the styrene. Most of this went into PS but copolymers such as ABS (Acrylonitrile-butadiene-styrene) and SAN (Styrene-acrylonitrile) took about one quarter of the styrene monomer.

GR-S was greatly inferior to natural rubber but particularly after the Korean War improved materials were produced. These now known as SBR (Styrene butadiene rubber) and take around a quarter of the world market for rubber depending chiefly on the relative prices of oil and natural rubber. Expanded PS is widely used as an insulating material accounting for about 1 million tons per annum.

Acrylic Polymers

Research on acrylics commenced with Dr. Otto Rohm in Darmstadt but commercial developments had to await an economic route to the synthesis of the monomer. Dr. John Crawford in 1932 discovered the acetone-cyanohydrin route to methyl methacrylate which turned MMA from a laboratory to a commodity chemical. In 1933 a sheet of polymethyl methacrylate (PMMA) was cast from MMA and "Perspex" was born.

Perspex was used during the war as aircraft glazing; the organic glass had now been found. PMMA is made under a variety of trade names, inter alia Plexiglas and Oroglas.

Many plastics were first used as substitutes for other materials; much of the impetus behind PVC was derived from the need in Hitler's Germany for a substitute or ersatz rubber and much of the obloquy of ersatz materials has rubbed off onto plastics in general. It is common to look down on plastics because they are substitutes for natural products. When a material like the melamine-formaldehyde "Formica" can be used instead of hardwood from the rain forests or when "parkesine" can replace ivory and thus reduce the incentive to kill elephants this view seems to be illogical.

In fact plastics are the most beautiful of materials. The Acrylic group on its own justifies this thesis although in all aesthetic judgements unanimity can not be expected.

Examples can be found in the fields of architecture, personal jewellery, engineering and building models, construction, furniture, sculpture and painting.

Acrylic glazing has permitted architects to design buildings free of the limitations of traditional materials and thus to enhance the built environment. Acrylics both alone and in conjunction with other materials are widely used by designers of furniture, jewellery and models. In the case of engineering models, these have a further advantage that acrylic models can be used to calculate stresses in the final construction.

"Perspex" has been used to make statues and other sculptures whilst acrylic paints are now the preferred medium for professional and amateur artists as well as internal decorators.

Nylon

1938 saw the first commercial use of a synthetic fibre - nylon. Initially used in toothbrushes, nylon stockings became legendary during the second world war and afterwards were worn by every woman in the developed world. Nylon became one of the World's staple fibres and also an important engineering plastic.

Nylon was one fruit of a decade of fundamental polymer research by Wallace Hume Carothers. Between leaving Harvard in 1927 and his untimely death in 1937 Carothers had distinguished between addition and condensation polymerisation; devised the molecular still to make very high mol wt polyesters; recognised that these superpolyesters (and polyamides) could be drawn into strong pliable fibres; contributed to the chemistry of acetylene and the commercial development of chloroprene. But for his untimely death a Nobel prize for Carothers would have been inevitable.

Polythene

No one will be unable to reply to the question "what happened on September 3rd 1939?" but only a few would say it was the date when polythene went into commercial production.

Polythene was however destined to have an important effect on the second world war. The Battle of Britain in 1940 was a significant turning point and the radar that helped to win that battle was critically dependent on polythene. Only seven years earlier polythene was "a waxy solid found in the reaction tube" to quote from the notebook of R.O. Gibson a research chemist at ICI Alkali Division. Polythene was an unexpected product of a reaction that had been intended to produce derivatives of benzaldehyde. This chance discovery was followed up and it was felt that polythene could replace gutta percha for the insulation of submarine telephone cables. Polythene did indeed provide excellent insulation for submarine cables and also for radar.
After the war polythene became available for less vital applications and soon replaced vitreous enamel for water

basins and buckets. At present millions of tons of polythene are used in domestic applications world wide. Polythene has revolutionized the packaging industry whether for 2 cwt sacks of fertilizer or cement or a bag for potato crisps. Polythene shopping bags have been welcomed as a convenience by housewives but what is more significant is that plastic containers serve to preserve foods world wide from pests and dirt so that the yield of agricultural products is increased in some cases by 100%.

Polythene is now widely used as an engineering polymer for example in moulded milk crates; for gas distribution piping it has completely replaced cast iron piping and it is widely used in chemical, water and other pipes. Polythene film is used in agriculture and building as well as for packaging, blow moulded polythene is used for bottles, toys and barrels.

There are many minor applications of polythene, low molecular weight polythene are waxes and polythene is widely used in spectroscopy. The powder is used as a sample support and polythene based filter, prisms and gratings were used to make the first Fourier Transform Infrared Spectra of polymers (3). Polythene is transparent in the Far infrared apart from an absorption near 70 cm.

As a result of this FTIR work it was possible to calculate the theoretical modulus of a polythene crystal and this was found to be very much higher than values obtained in practice (4). It was clear that there was great scope for making very strong polythene fibres. Severe problems in the extrusion of the fibres had to be overcome. Workers at the University of Leeds solved these problems and high strength PE fibres are now on the market. Another solution to the extrusion problem was found by the Dutch State Mines who also market high performance polythene (HPPE) fibres.

The DSM HPPE fibres have a high specific strength (2.65 GPa), high modulus (87 GPa) and low density (0.97). If we compare free breaking length (the length of rope that breaks under its own weight) HPPE is 10 times as strong as steel. HPPE has a free breaking length of 336 km compared to 193 (Aramids), 171 (carbon), 92 (Nylon), 76 (Glass) and 37 (Steel) for other ropes. Composites reinforced with HPPE have excellent impact strength whilst hybrid carbon fibre/HPPE composites give the highest energy absorption for tubes (110 kJ/kg) (5).

Worldwide the annual production of polythene approaches 30 million tons making it the largest tonnage polymer.

The ICI process used very high pressure to produce a polythene of low density (0.92). Research on metal organic catalysts by Karl Ziegler led to a low pressure process producing medium density polythene (0.94). Phillips, Standard Oil and other processes enable high density (0.96) polythene to be produced. Newer types of polythene are now available of which LLDPE (linear low density PE) is the most important.

Polypropylene

Using a Ziegler type catalyst Giulio Natta in 1954 in Milan polymerized propylene and deduced the structures of the various isomers that became possible with the addition of a methyl group to the polythene chain. He also copolymerized ethylene and propylene to give new elastomers; for this he shared a Nobel Prize for Chemistry with Ziegler.

The production of polypropylene (PP) grew faster than any other plastic and it is now only surpassed by PE and PVC. Global production is around 10 million tons.
PP is widely used as a fibre and in engineering. Broadly speaking applications for PP are the same as PE except where the poor oxidation stability of PP rules it out.

Polyvinyl chloride PVC

Decomposition products of PVC may be detected as low as 70°C, yet the amount of PVC used is exceeded only by the polyolefins. PVC world capacity is 20 million tons. This is largely due to advances in stabilizers and the use of plasticisers that enable PVC to be processed at moderate temperatures.

Vinyl Chloride was polymerized using U.V. by Baumann in 1872. PVC is one of the most versatile of polymers, ranging from large water pipes, and roofing, to wine bottles, toys, fashion garments and electrical insulation. PVC is widely used in packaging, records and tapes. PVC has replaced rubber as insulation for low frequency electrical cables, and PVC bottles are used for mineral waters, cheap wines and soft drinks. Clothing from nappies to high fashion use PVC. Household gloves, hand grips, floor coverings, gramophone discs, wallpaper, leathercloth and leather substitutes

all use PVC. Large tonnages are used in piping and building. Unplasticised PVC (UPVC) is the material of choice for window frames and guttering.
The most important fluoro polymer Poly Tetra Fluoro Ethylene (PTFE) was discovered in 1938 (6), although I.G. Farbenindustrie had patented PCTFE in 1934 (7).

PTFE is chemically inert, has exceptional resistance to weathering and heat, excellent electrical insulation, a very low coefficient of friction and adhesion.

Non-stick frying pans and faster skis are the popular manifestations of PTFE but it has many chemical and aeronautical uses in chemical and aeronautical engineering where the high costs are acceptable. Other fluoropolymers are used as seals, roof coverings and for electrical applications.

Polyurethanes

In 1937 a German patent on polyurethane was granted to Otto Bayer of I.G. Farbenindustrie as part of the drive in Germany for an ersatz rubber. Initially Bayer's work was regarded as useless but the possibilities of PU foam were later realized and patented in 1941 and 1942. Currently Bayer, one of the companies formed on the break up of I.G., now make approximately 4 million tons of PU per annum. Polyurethanes are found as elastomers, moulded plastics, fibres, coatings and foams; only silicones have such an extensive range and their production is only 10% that of PU. PU are used in a new moulding technique of "Reaction Injection Moulding" largely due to work by the Dow company. When reinforced with long glass fibres Reinforced Reaction Injection Mouldings (RRIM) provide very strong materials with short production cycles and find wide application in automobiles and buildings especially those needed in a hurry.

PU foam provides thermal insulation in refrigeration, chemical plant and buildings. Domestic freezers and refrigerators would not be economically viable without PU foam. Hard-wearing coatings on aircraft, trains, boats and commercial buildings are usually based on polyurethanes (8).

In 1953 the Nobel prize was awarded to Herman Staudinger, Hermann Schnell patented polycarbonates and the first epoxy resin insulated current transformer went into commercial service in Redcar, North Yorkshire.

Polycarbonates are tough and transparent with good electrical insulation and low flammability. Most applications are in the electrical industry but glazing and antivandalism are large markets. There is capacity mainly in Germany and the U.S. for half a million tons. There has always been a close connection between plastics and the electrical industry dating back to Sir James Swinburne. Alkyl polyester resins were first made in the laboratories of General Electric and the use of epoxy resins revolutionized the design of electrical equipment (9). The replacement of the flammable, liquid hydrocarbon oil by a solid epoxy resin meant that the location of transformers was no longer constrained by the need for an oil tank. In addition, epoxy insulated material required minimal protection out-of-doors.

<u>Composites</u>

Wood fibre (saw dust), cotton and various other fibres were used to reinforce phenol formaldehyde resins but the high pressure compression moulding process precluded the use of longer fibres. Polymers that cured by an addition rather than a condensation reaction could be used at low pressure, and thus longer fibres usually of glass are widely used to reinforce polyester, epoxy and urethane resins.

The discovery of carbon fibres at the Royal Aircraft Establishment in Farnborough was truly a milestone as carbon fibre reinforced epoxy resins enabled improvements in strength and modulus of an order of magnitude to be made.

CF Composites have revolutionized the design of aircraft, racing cars and sports equipment, especially tennis racquets and fishing rods. Amongst numerous advances in medicine CF/acrylic composites have eliminated breakages in dentures (10).

In 1942 GRP (glass reinforced polyester) was introduced. GRP has now replaced wood in boat building initially in dinghies, then yachts and now for minesweepers. GRP is widely used for roofing, pipes, motor cars, temporary huts, water towers, street furniture and many other applications in construction, transport and agriculture. Glass reinforced nylon was introduced in 1946 and is now an important "engineering plastic" providing high mechanical strength with electrical insulation. Other thermoplastics, polystyrene, polypropylene, acrylates and terephthalates are available with glass reinforcement.

The success of GRP and CF/Epoxy has led to the emergence of reinforced plastics as a separate group of materials known as "composites". The term is imprecise; most writers restrict the term to long fibre reinforced addition compounds, but in fact any filled polymer may be regarded as a composite. Indeed dentists use the term composite in that, etymologically correct, way.

Aramid fibres have also been used to make high strength composites. These fibres are liquid crystal plastics. Under shear (e.g. in extruders) L.C. plastics become more regular with a concomitant increase in strength. There is considerable current interest in L.C. plastics because of this increased order and strength. L.C. plastics do not break as clearly as metals or other polymers but fail in stages rather like wood, which can provide an extra factor of safety.

Another current area of interest is conducting polymers. Plastics are traditionally insulators but conducting polymers are now being made and tested for use, e.g. in specialised batteries. The goal of polymer transistors is in sight.

Conclusion

In the last 150 years the synthetic plastics industry has been born and has grown to out-rank much older metals. It is hoped that this survey, although eclectic, may convey some flavour of this exciting and vital industry.

References

1) Landmarks of the plastics industry 1862-1962 Imperial Chemical Industries. This book from I.C.I. Plastics Division has been drawn on passim

2) B.P. 16,274

3) T.R. Manley and D.A. Williams, IUPAC Macromol. Divn. A 529 (Prague 1965)

4) T.R. Manley and C.G. Martin, Polymer 14 491 (1973)

5) DSM Research 6160 MD Geleen, Netherlands

6) U.S.P. 2,230,654

7) B.P. 465,520

8) The Bayer Story 1863-1988, E. Verg, Bayer AG. Leverkusen.

9) T.R. Manley, K. Rothwell and W. Gray, Proc. I.E.E 107A,213 (1960)

10) A.J. Bowman and T.R. Manley, British Dental J. 156,87 (1984)

Energy

Fossil Fuels 1850 to 2000

Ian Fells

DEPARTMENT OF CHEMICAL PROCESS ENGINEERING, UNIVERSITY OF NEWCASTLE UPON TYNE, NEWCASTLE UPON TYNE, NE1 7RU, UK

1 INTRODUCTION

Civilisation began when man discovered how to make and control fire. The first fuel was probably wood and even today fuel wood provides just over ten per cent of man's energy requirements. Nearly eighty per cent is produced by burning fossil fuels; oil, coal, gas and associated hydrocarbons. They are a non-renewable resource and took 200 million years to form. Man started to burn them around 1000 years ago when simple coal mining techniques were developed, although seepages of oil and tar had provided a primitive fuel many hundreds of years earlier. Once the Industrial Revolution got under way towards the end of the 18th century, the rate of consumption increased. We now know that in an economy undergoing industrialisation an increase of one per cent in gross domestic product (GDP) requires a 1.5 per cent increase in energy use (known as the energy coeft).

The earliest fuel consumption figures available[1] relate to the UK when in 1800, the UK energy coeft. was just about 1.5. In the latter years of the 20th century this figure has fallen to 0.5 in the UK and the rest of Western Europe and is typical of mature industrialised economies. In rapidly expanding and industrialising economies such as those around the Pacific Rim the energy coeft. is still 1.5 and with some economies growing at between 7 and 8 per cent the increasing demand for fossil fuels in these developing countries could well dominate world energy demand early next century. Matters are exacerbated by massive increases in world population. Numbers have risen from 1.7 billion in 1900 to 5 billion today, and will rise to 10 billion in 2060. Each additional person requires his or her energy ration and anticipates a steadily improving life style. All these pressures conspire to increase the world's energy demand and use.

2 ENERGY RESERVES AND USE

This presents two problems. First, what is the extent of the world's reserves of fossil fuels; how long will they last? Second, all fossil fuels burn to form carbon dioxide; each tonne of oil burnt produces 3.3 tonnes of carbon dioxide, for example. Carbon dioxide has been steadily building up its concentration in the atmosphere since the start of the Industrial Revolution and has risen from 270 parts per million (ppm) in 1700 to 350 ppm today. Together with methane, to a lesser extent, and other gases this increase in concentration gives rise to the "greenhouse" effect which is perceived as a serious threat to the environment, causing global warming and destabilisation of the weather machine. Sulphur and nitrogen oxide emissions from coal and oil combustion cause acid rain which seriously damages forests, crops, buildings and health.

The finite nature of the world's fossil fuel resources, taken in conjunction with the destructive effect on the environment caused by their combustion, makes an unassailable case for conserving and constraining their use. On the other hand, the expectations of the rapidly rising world population require ever increasing supplies of energy to 'fuel' their aspirations. This presents the governments of the world with a political as well as a resource dilemma which will steadily get worse. The extent of world fossil fuel reserves is always a matter of judgement and figures, particularly in the oil industry, have sometimes been distorted for commercial reasons. The best and most objective figures are those provided by the World Energy Council and published in 1989. Reserves and consumption figures are set out below in terms of tonnes of oil equivalent (TOE). One TOE equals 42 Giga Joules (GJ). Figures are divided between industrialised countries (IC) and developing countries (DC) and given in Table 1.

Table 1

WORLD ENERGY RESERVES AND CONSUMPTION (1987)

	IC	DC	World
Economic reserves (btoe)			
Coal	380	160	540
Oil	16	100	116
Gas	50	46	96
Total fossil fuel	**446**	**306**	**752**
Uranium, non-communist world:			
Thermal reactors (LWRs)			33
Fast reactors			>2000
Consumption (mtoe)			
Coal	1200	1190	2390
Oil	2060	880	2940
Gas	1250	300	1550
Total fossil fuel	**4510**	**2370**	**6880**
Nuclear	370	40	410
Hydro/geo.	330	190	520
Wood	100	1100	1200
Total (btoe)	**5.3**	**3.7**	**9.0**
toe/cap	**4.8**	**0.9**	**1.8**

Table 2

Reserves/current consumption (yr)

	IC	DC	World
Coal	–	–	225
Oil	–	–	40
Gas	–	–	60
Thermal reactors	–	–	80

The figures given here refer to proved recoverable reserves. Estimates of proved and probable reserves are sometimes quoted but should be accepted with caution.

A simplistic life time for the different fossil fuels can be obtained by dividing the reserves by current annual consumption. The results are given in Table 2. Figures for uranium are included as it is a fossil fuel, although not always thought of as such. There is a striking difference in uranium reserve potential compared with other fuels if its energy release is calculated, assuming it is used in the highly efficient breeder reactor where the abundant uranium 238 isotope is converted to plutonium for use as a fissionable fuel rather than merely burning up the uranium 235 isotope in the currently commercial thermal reactors. The multiplying factor is 60.

3 ENERGY DEMAND

Of course these "lifetimes" shorten dramatically if, as seems likely, annual consumption rates increase. The attempts to predict future energy demand made by three well known and broadly based foundations, the International Institute for Applied Systems Analysis (IIASA), the World Energy Council (WEC), and the World Resources Institute (WRI) have been compiled by AEA Technology and are shown in Table 3.

There are noticeable differences in the predictions for 2020. IIASA is pessimistic about the contribution of renewables (8 per cent) and optimistic about the future nuclear component (21 per cent), both WEC and WRI give renewables a 20 per cent share whereas nuclear is 12 per cent for WEC and a low 5 per cent for WRI. In any event the brunt of demand will be taken by the fossil fuels coal, oil and gas; all the scenarios, whether high or low, give a figure of between 68 and 71 per cent. WRI is more enthusiastic about gas and less enthusiastic about coal, probably for environmental reasons. All predict an increase in energy demand over the 1987 figure of 9 btoe ranging from 46 per cent IIASA, 11 per cent or 67 per cent WEC to 18 per cent WRI except for the WRI "low" scenario which is not a projection at all but a "target" involving enormous improvement in the efficiency of energy use.

Table 3

SCENARIOS BY FUEL FOR 2020

	IIASA (1981) btoe	%	WEC (1986) btoe	%	WRI (1988) btoe	%
Coal High	4.6	28	6.0	31	1.4	13
Low			4.6	30	1.4	17
Oil	4.4	27	4.4	23	3.1	29
			3.2	21	2.3	29
Gas	2.7	16	3.4	17	3.1	29
			2.6	17	2.3	29
Nuclear	3.5	21	2.4	12	0.5	5
			1.7	12	0.5	6
Renewables:	1.3	8	3.4	17	2.6	24
Biomass etc			3.0	20	1.5	19
Total	**16.6**	**100**	**19.6**	**100**	**10.7**	**100**
			15.1	**100**	**8.0**	**100**

These predictions were made over the period 1981 (IIASA) to 1987 (WEC) and 1988 (WRI). More recently (1989) the WEC has published "Global Energy Perspectives 2000 - 2020". In it the figure for world energy consumption between 1985 and 2020 has been revised to grow between 50 and 70 per cent. The rise in demand will be uneven with spectacular rises in developing countries and particularly Centrally Planned Asia.

Again there will be wide differences in per capita consumption, although the world average will stay at around 1.6 toe. In 1985 average per capita consumption in the South, including non-commercial energy sources such as fuel wood stood at 0.65 toe as against 4.25 toe in the North. By 2020 the corresponding figures will be 0.8/0.9 toe in the South and 4.45/5.15 toe in the North, so there is hardly any improvement in the ratio for the South in terms of consumption. The situation is particularly bad in countries such as Sub-Saharan Africa and South Asia where per capita consumption will rise from 0.36 toe in 1985 to only 0.39/0.46 toe in 2020. Of this, non-commercial energy would still constitute between 30 and 45 per cent of demand.

It is clear that chronic poverty in the energy field will continue to exist in a region whose population is expected to double from 1.4 bn to 2.8 bn by 2020; that is, one third of the world population.

4 ENERGY SUPPLY

On the fuel supply side the 1989 WEC predictions are that coal demand will rise after 2000 to between 3 and 4 btoe but that before 2000 there will be strong competition from hydrocarbons, particularly gas. Nevertheless coal will only have around 30 per cent of the market in 2020. Outlets are restricted and environmental constraints constitute a check on development.

Natural gas, because of its environmental advantages in not containing sulphur and producing less carbon dioxide per unit of heat output than coal or oil, is better placed than oil to maintain its share of world demand at current levels.

The nuclear power sector, despite its environmental advantage in producing neither carbon dioxide nor acid rain, suffers from public acceptability problems since Chernobyl and also increasing financial difficulties, so the WEC (1989) report has scaled down the anticipated nuclear contribution to between 7 and 8 per cent compared with 4 per cent in 1985.

Non-commercial energy sources will increase in the South and will still provide between 15 and 25 per cent of Third World energy needs in 2020 as compared with 33 per cent in 1985. The inevitable consequence is increased pressure on agriculture, society and the environment.

Overall, Renewable Energy Sources, that is hydropower and non-commercial sources taken together, are unlikely to provide more than 20 per cent of energy demand by 2020, despite the enthusiasm of their supporters.

The position of oil will be maintained to a higher degree than foreseen in earlier reports and will still be meeting 26/28 per cent of demand by 2020 compared with 32 per cent in 1985. In particular, Third World demand will rise from less than 0.7 btoe in 1985 to 1.4/1.6 btoe in 2020 and the South's share of world consumption will increase from 26 per cent to 43/44 per cent in 2020 whilst industrialised countries will have stabilised their demand at 1985 levels.

5 WORLD TRADE IN ENERGY

Oil is the largest single commodity traded in the world today. In 1989 trade rose by 6 per cent over the 1986 figure to 28.5 million barrels/day (b/d) (7.2 bls of oil = 1 tonne). OPEC alone obtained net export revenues of $117 bn in 1989. The US is the single largest importer of crude oil and in 1989 imported 7.8 million b/d, an increase of 19 per cent on the 1988 figure. Figure 1 shows the distribution of oil reserves in different countries. The role of OPEC is crucial to the continuous supply of oil in the world, particularly to the industrialised countries, it is ironic that OPEC was formed in 1960 because of the intransigence and greed of the large international oil companies, particularly Exxon. The fact that the major reserves of oil lie in a politically highly unstable area and that oil looks set to provide the major fraction of world energy into the next century, and certainly past 2000, bodes ill for price and supply stability.

The rising fortunes of natural gas with its environmental advantages has led to a broad geographic spread of gas supplies. Trade in natural gas is to some extent constrained by the development of gas pipe lines although a network of pipelines in the North Sea and from Russia into Europe are rapidly extending trade across national boundaries. World-wide, ten countries export to 25 others and eight nations export liquified natural gas. Once again world natural gas reserves are concentrated in particular locations; USSR and the Middle East between them have 67 per cent of world reserves, Western Europe an important 5 per cent. Of the world's ten largest gas fields, five are in USSR and two in Western Europe. As far as Europe is concerned a gas market controlled by Norway, Russia and possibly Holland is a strong possibility. A better arrangement would be the growth of a 'spot'

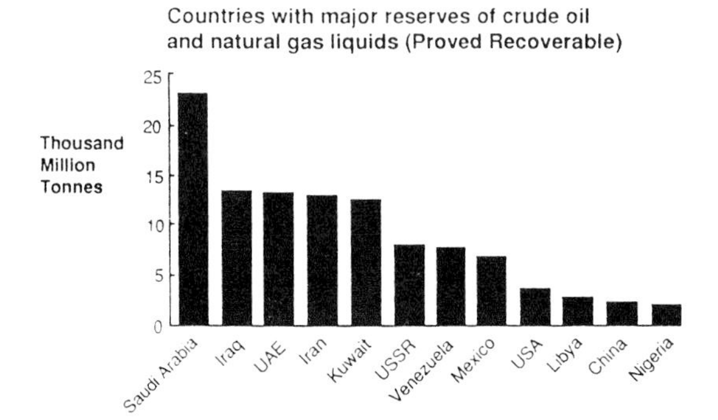
Countries with major reserves of crude oil
and natural gas liquids (Proved Recoverable)
Thousand
Million
Tonnes
25
20
15
10
5
0
Saudi Arabia
Iraq
UAE
Iran
Kuwait
USSR
Venezuela
Mexico
USA
Libya
China
Nigeria

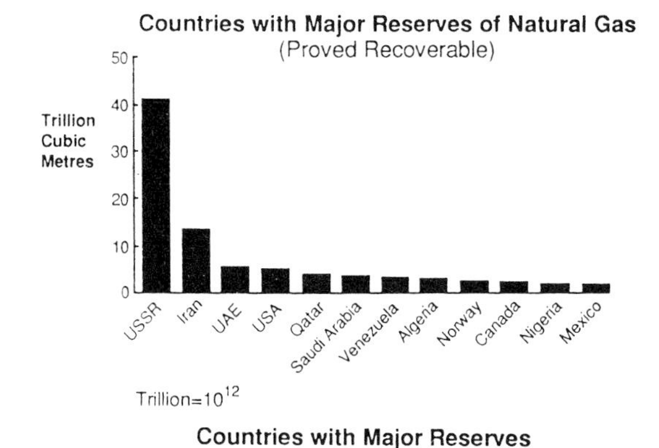
Countries with Major Reserves of Natural Gas
(Proved Recoverable)
Trillion
Cubic
Metres
50
40
30
20
10
0
USSR
Iran
UAE
USA
Qatar
Saudi Arabia
Venezuela
Algeria
Norway
Canada
Nigeria
Mexico
Trillion=10^12

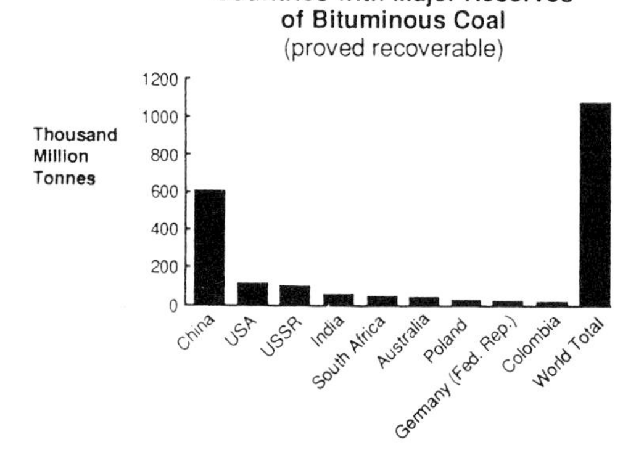
Countries with Major Reserves
of Bituminous Coal
(proved recoverable)
Thousand
Million
Tonnes
1200
1000
800
600
400
200
0
China
USA
USSR
India
South Africa
Australia
Poland
Germany (Fed. Rep.)
Colombia
World Total

Figure 1

market in gas which could constrain prices, although they will inevitably rise, possibly quite quickly in the late 1990s.

About 360 mt of coal is traded world-wide; that is, about 11 per cent of production. Australia is the largest exporter with the US not far behind, followed by South Africa. China has great but unrealised coal export potential. Coal's major competitor in the world market is now natural gas and this will constrain growth in the coal trade to around 2 per cent for a time until the price of natural gas gives coal an advantage when burnt using clean coal technology.

6 TECHNOLOGY AND CHEMISTRY

Perhaps the most important chemical reaction that exists is the combustion of carbon-based materials in air to form water and carbon dioxide with the generation of heat. Wood, coal, oil and gas can all be easily burned and have provided first heat and later electricity via heat engines. Combustion provides a reliable and continuous source of energy which makes our complex civilisation possible. Without heat and electricity we spiral down into chaos within as little as 24 hours, leaving people stranded in lifts, hospital life support systems paralysed and intolerable conditions of cold or overheat as heating and air conditioning systems fail.

As is often the case, people used fire and later boilers and furnaces before the chemistry of combustion was understood. After Priestley and Lavoisier, the work of Humphrey Davy at the Royal Institution on the combustion of methane following the Felling Colliery disaster in 1812 (ironically a disaster involving the extraction of another fuel, coal) started a train of combustion research by chemists that continues to this day. A great industry is built round research into combustion in gas turbines, fluidised bed combustors, furnaces, internal combustion engines and combined heat and power schemes, all with the intention of improving our understanding of the complex combustion process with its chain reactions and free radical mechanisms. But also with the intention of improving efficiency of fuel usage. Indeed the whole development of chemistry has been strongly influenced by fuel and combustion.

The treatment of fuels once they have been extracted from the earth involves a wide range of chemical techniques. Originally the primary fuel was coal; in the UK alone 287 mt were mined in 1913 and 100 mt exported. The Navy was only just in the process of switching to oil. The chemistry of coal utilisation included the analysis of coal, both proximate and ultimate, as well as many years of work in trying to establish the structure of the coal molecule. This information, embodying the measurement of calorific value, volatile content, propensity to cake during heating and so on was included in several methods for the classification of coal culminating in the NCB (National Coal Board) classification used today. This enables different coals to be directed to particular uses as steam raising coals, coal suitable for conversion into strong metallurgical coke or, more appropriate, for gasification and the production of smokeless fuels. The techniques of carbonisation and gasification are central to the development of coal technology. Coal is carbonised with the main object of converting it into products which can be utilised with improved efficiency and recovering valuable by-products which are generally lost when raw coal is consumed. An important product is coal gas and in the late 18th century coal gas was produced for illumination and this role of gas continued until well into the 20th century. Gas gradually came to be used as a clean and convenient fuel in the latter part of the 19th century. The by-products of coal gas manufacture were active coke which could itself be used as a clean, solid fuel and a host of organic chemicals, such as phenol, and oils which could themselves be converted into fuels, including petrol, and spawned an important chemical industry. This role was only taken over by the petrochemical industry, based on oil, in the late 1930s despite the discovery of oil early in the 19th century, when Col. Drake drilled the first oil well in Texas (although he was looking for water at the time).

The chemistry and chemical engineering required to process oil had originally been developed in Scotland in the early 19th century by Young who retorted oil shale (Kerogen rock) available in the Bathgate area and then went on to distil and chemically treat the products. As the oil industry grew, a powerful group of techniques, including catalytic and thermal cracking, reforming, acid treatment and so on, brought chemistry to bear and the petrochemical industry was born, leading to polymers, fibres, paints and a host of essential chemical products. Rather later, around the mid 1950s, steam reforming of naphtha led to a process to produce town's gas which replaced coal gas and this was only superseded in the UK when natural gas was discovered under the south North Sea in 1967. Natural gas brings its own chemical problems; the formation of clathrate compounds with water and the removal of sulphur compounds from sour gas.

Chemistry and the fuel industry have both benefited from their symbiotic relationship.

7 ENVIRONMENT

The development of the oil industry and, more particularly, the coal and coking industries, were carried out initially with little thought for the depredation and dilapidation they caused to the environment. In the UK the clean air act of 1956, the result of the disastrous "smogs" in previous years, triggered the national conscience into a belated interest in cleaning up the environment, but industry was reluctant to embark on an expensive clean up campaign without legislation. Even now it has required the European Community (EC) to adopt the Large Combustion Plants Directive (1988) and other legislation concerning dumping of waste and so on to force industry and others to clean up our polluted environment. Problems outside the EC are highlighted by countries like Poland, Czechoslovakia and East Germany where the environment has been seriously, perhaps irrevocably damaged. The fuel industries must shoulder a good deal of the blame for global as well as national pollution. The production of acid rain by burning coal and oil on a large scale for electricity generation, leading to the destruction of forests and agriculture, as well as sterilising hundreds of thousands of lakes, is only dwarfed by the damage caused by the production of carbon dioxide which leads to global warming via the "greenhouse" effect. The consequences of this are as yet unknown, but the prospects of rising sea levels coupled with destabilisation of the weather machine are daunting. Most of this carbon dioxide is the result of burning one fossil fuel or another as Figure 2 shows.

Whilst chemistry can do something to alleviate the formation of acid rain by scrubbing out the sulphur and nitrogen oxides from the flue gases, the prospect

Sources of man-made carbon dioxide emissions 1980-85

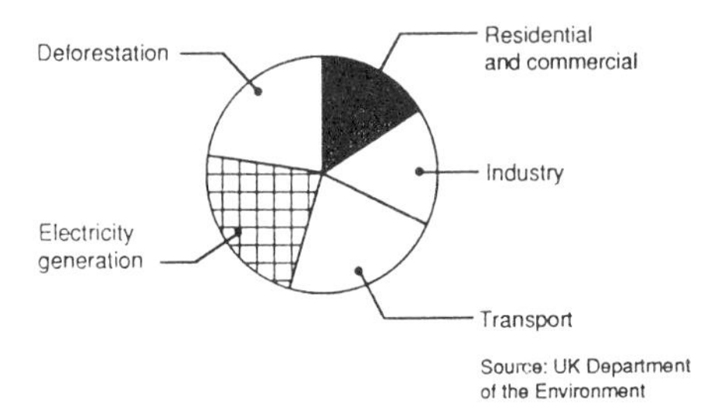

Carbon emissions from fossil fuel use by region 1987

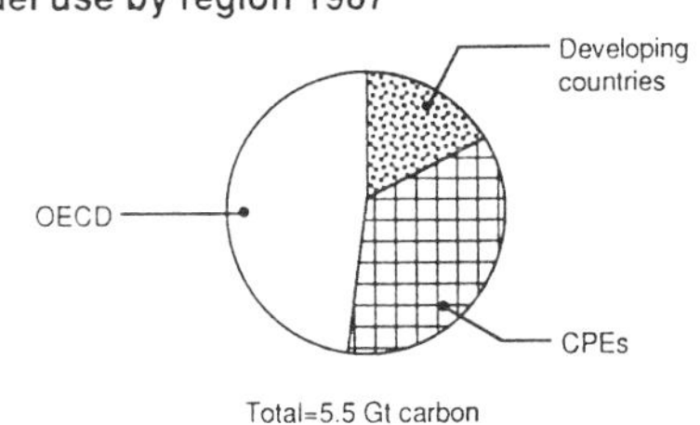

Figure 2

Cutting the UK's carbon dioxide emissions

Contribution towards target

Source: Energy Technology Support Unit, UKAEA, 1989

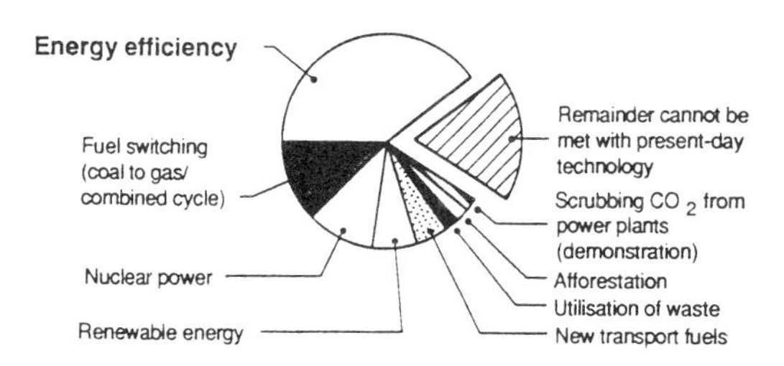

Target: 50 per cent reduction in 30 years

Figure 3

of removing carbon dioxide and then disposing of it in some acceptable way has proved an insuperable problem. Figure 3 shows some of the techniques which could be used to reduce carbon dioxide emissions in the UK, but the cost would be considerable.

8 CONCLUSION

Fossil fuels have sustained the growth of civilisation for several thousand years but the accelerating demand for more and more energy, encouraged by rapidly improving life styles and exacerbated by rapidly rising populations, particularly in the South, are beginning to emphasise the finite nature of fossil fuel reserves. Nevertheless, forward prediction suggeststhat fossil fuels will still supply some seventy per cent of our ever increasing energy demand well into the next century. More efficient use of energy is a paramount priority but even then the increasing combustion of fossil fuels leading to environmental damage by acid rain and, more insidiously, by carbon dioxide via the "greenhouse" effect, point to an uneasy and uncomfortable compromise between environmental protection and economic growth. Without population control and downgrading of lifestyle expectations, nuclear power will remorselessly take over from fossil fuels as reserves become depleted and world energy demand continues to accelerate.

References

1. "Energy, implications for economic growth", Ian Fells, Chemistry & Industry, Vol 6 1980, p 655-658

Data has been obtained from the following sources:

World Energy Council publications, 1989 Survey of Energy Resources.

BP Statistical Review of World Energy

Shell Briefing Service Publications, "Global Climate Change" 1990, 'Prospects for Natural Gas" 1990

Data compiled by AEA Technology from projections by Institute for Applied Systems Analysis 1981, World Energy Council 1986, World Resources Institute 1988

Explosives in the Service of Man

J.E. Dolan

FEDERATION OF EUROPEAN EXPLOSIVES MANUFACTURERS, CEFIC, 250 AVENUE LOUISE, BRUSSELS, BELGIUM

EXPLOSIVES IN THE SERVICE OF MAN

We live in an age of technology the key to which is power. Power is so essential to our civilisation that it is, perversely, taken for granted and with little realistic consideration of the consequences of what would happen if it were not available.

The ability of this power dependent era to progress is directly related to the ability of science to find and exploit new sources of power. Indeed civilization as we know it could not survive if deprived of electricity with which to power industry and home and of oil and petrol for transport. Cities without power would die within the week and their citizens with them. Little wonder then that so much of scientific and engineering research has been spent in prospecting for new resources from which to tap this all important power. Over the centuries power has been obtained from the movement of water, from steam, from petrol and oil, from coal, from chemicals and other fuels, and finally from the very heart of the matter itself, in atomic power. The history of the way in which power has been developed is a fascinating story and, in the array of power sources that are available today, that obtainable from EXPLOSIVES is and will continue to be, one of the most spectacular.

It is, however, a sobering and intriguing thought that this most sophisticated form of energy is fundamentally the same as the oldest form of energy known to man: - FIRE.

The chemistry of the modern high explosive can, in truth, lay claim to be fundamentally the same as that of the first chemical experiment ever carried out by thinking man when in the darkness of his caves in the dawn of prehistory, he first struck flint upon flint and created fire. Fire which was so vital to his continued existence and to his slow progress towards civilisation that it was worshipped as Divine for countless centuries.

This chemical combustion reaction gave to these first unknowing experimenters the key to chemical energy: the combination of fuel and oxygen to release energy in the form of heat and light. Light with which to illuminate the darkness of his caves and heat with which to insulate himself from the alien climate in which he lived and enable him to survive in a hostile world. This was the crucible of fire in which was born the first stirrings of civilisation on this planet.

Let us then spend some little time and examine this most important of the energy producing reactions in detail and trace how FIRE can be made to become EXPLOSIVE.

There are only two things that are necessary for a combustion reaction - fuel and oxygen.

The secret of explosive technology is to take this inter-molecular reaction and speed it up to the point where it is taking place faster than the speed of sound.

Today everyone is aware of the fact that any object that travels faster than the speed of sound is accompanied by a supersonic shock wave. This applies as much to chemical reactions as to supersonic jet aircraft. It is the shock wave associated with high speed chemical reactions that is the target of the explosive technologist.

The art of the explosives technologist therefore lies in taking this simple combustion reaction and in controlling the speed with which the energy and gases are produced. This in turn, is very largely a question of the way in which the oxygen and the fuel are brought together. The normal combustion reactions with which we are familiar take place by bringing controlled quantities of fuel into contact with an excess of oxygen and burning it at the contact surface. If, however, the oxygen is intimately mixed with the fuel in oxygen balanced proportions the reaction takes place in the mass at a rate very close to the velocity of sound and therefore becomes a quasi detonation.

The first step towards an explosives technology was taken thousands of years ago when it was accidentally discovered that certain solid materials could be substituted for atmospheric oxygen. This discovery was made without any knowledge of the chemistry involved when the Chinese, reputedly, invented BLACKPOWDER.

Over the years many solid chemicals have been found to have this oxygen carrying property, the classic in early historical use being Chile Saltpetre (Potassium Nitrate).

These oxygen carrying chemicals, principally the nitrates, when mixed with such fuels as charcoal, sawdust or wheatflour, enable combustion reactions to be carried out by solid mixtures. The main characteristic of such solid mixtures is that, containing their own source of oxygen, the combustion reaction occurs even when atmospheric oxygen is excluded from the reaction.

The best known of all such solid combusting mixtures is, of course, Blackpowder or Gunpowder. Blackpowder consists of a mixture of Charcoal, Potassium Nitrate and Sulphur. These ingredients are intimately mixed together by milling to give a fine black powder which serves admirably to define the second important step in increasing the speed of a combustion reaction. This step relies on the fact that the speed of any chemical reaction is dependent on the pressure under which that reaction takes place. The higher the pressure the greater the reaction speed.

When small quantities of blackpowder are ignited in the open and under atmospheric pressure the reaction, whilst both energetic and fast, is in no way explosive. If, however, the same blackpowder is confined in a solid shell which does not allow the combustion gases to escape the gases build up a pressure inside the shell. This increasing pressure cumulatively increases the speed of the combustion until finally the speed of the flame front passes the threshold of sonic velocity in the material, shock wave conditions are established and the reaction becomes a detonation.

Pressure dependent inter-molecular explosive such as Blackpowder are called DEFLAGRATION EXPLOSIVES.

The origins of gunpowder or blackpowder are lost in the mists of time. It is reputed to have been known to the Chinese over 2,000 years ago. In spite of this very long history, there is no recorded use of Blackpowder in practical mining or quarrying before the seventh century. Before then it had a varied and mystical career in which the religious, warlike and the entertaining were all intrinsically mixed up. Until about the tenth century there was no clear idea about the correct proportions of nitrate, charcoal and sulphur to use, and the preparation was surrounded in mysticism. The association of charcoal and sulphur with the evil spirits and the satanic, together with the ancient rites of fire worship sustained the association with the mystical and led to the widespread use of Fireworks to dispel evil forces, a practice which is still very much a part of life in China today.

The first reference to the use of Blackpowder for practical purposes is in the military field where Marcus Greacus in the year 700 A.D. describes its use in crude rockets, in thunder flashes for demoralizing the enemy and in Greek Fire. There is, in this military use, a possible commentary on human nature in that civilised man's first real use of major discoveries would seem to be for destructive rather than for constructive purposes.

The military application continued to dominate the development in Europe from the 13th century when the Franciscan monk, Roger Bacon formulated what was to become the basic composition for propellant gunpowder. It is interesting that Bacon considered this knowledge liable to such misuse that he concealed it in an anagram which was not completely broken until 1904. But, in fact, by the 15th century an industry had grown around blackpowder for military application but still no use had been made for mining purposes. This had to wait for a further two centuries. It was, in fact, first used for mining in Hungary in the 17th century. The potential was immediately recognised and the use soon spread to Germany and Britain.

The development into use for mining may now seem

obvious but, at the time, it was a brilliant piece of innovative engineering which has, in the succeeding centuries revolutionised mining, quarrying and civil engineering practice and undoubtedly significantly increased the rate of scientific progress.

The earliest attempts at using Gunpowder for blasting purposes consisted of simply pouring the powder into the naturally occurring fissures and breaks in the rock. The idea of preparing special shotholes followed very quickly and were recorded in use by 1637. These early shotholes were made with iron tipped borers about 10 cms. in diameter driven in by jack hammers and the holes were closed with wooden wedges called "shooting plugs" in an attempt to retain the gases and develop the explosive effect.

During the 18th and first part of the 19th century firing of gunpowder shots was a somewhat risky procedure. A charge of gunpowder was poured into the shot-hole and a long needle or stick inserted into the charge. The bottom of the stick was in the charge and the top protruded from the top of the shot-hole. After filling the rest of the shothole with clay stemmimg (to seal in the gases) the stick was withdrawn and the hole so left was carefully filled with loose, finegrained, gunpowder. A piece of touch paper (which was supposed to take about half a minute to burn but seldom did) was lit and laid on the gunpowder trail. The final, and from the shot-firers point of view, the most important act of the performance was for the shot-firer to run as fast as possible.

This original crude method was vastly improved in 1831 when William Bickford of Cornwall invented and patented "Miners Safety Fuse" with its reliable burning speed. Safety Fuse consists of a core of Blackpowder encased in a textile sheath which allows the gases from the combustion to escape so that there is no build up of pressure. It therefore burns at a regular speed, in this case at a rate of one centimeter every second.

The invention of safety fuse enabled the introduction of a known delay between igniting the fuse and the detonation of the charge. It allowed a further, and quickly discovered refinement in that sequential firing could be achieved by using a combination of different lengths of safety fuse.

The final refinement of the system of blasting using Blackpowder was achieved when electrical firing was introduced. The idea of using an electric spark to set off blackpowder had occurred to Benjamin Franklin as early as 1751, and the safety aspect of this method of firing was so attractive that several people attempted to develop the idea. The first demonstration of a practical system was delayed until 1804 when Baron Chastel of Austria showed that a series of military mines could be ignited by electric sparks. At this point modern technology steps into the picture for the first time. Electric firing became a practical procedure by about 1835. This facility coupled with safety fuse sequential firing introduced the rudiments of explosives engineering and by 1850 rounds, particularly in underground mining, were being scientifically designed and the modern science of explosives was born.

We therefore see that, prior to 1850, man's use of chemical energy to reduce the physical work load associated with mining, civil engineering and such like activities, was limited to those tasks which could be reliably performed by Blackpowder in its various forms.

This was the situation in 1850 when the third and final step towards the truly supersonic chemical reaction was achieved. The scientific position at this time was that chemical combustion had been developed to the point where it was independent of atmospheric oxygen and capable of being booted to supersonic velocities by means of confinement and thereby produce an explosion.

The next step was to find a technique whereby the high speed regime could be sustained without the necessity for artificial confinement. This step was accomplished by the introduction of the intra-molecular reaction, that is, one in which the necessary constituents for combustion are contained within the one single chemical substance and are not separate chemicals.

That this step was achieved is almost entirely due to the inventiveness of one man, Alfred Nobel. It began with the discovery of Nitroglycerine. Nitroglycerine was discovered by the Italian chemist Sobrero in 1860 during experiments into the nitration of organic compounds in an attempt to synthesize urea. One of the organic substances he tried was glycerine.

In this experiment he mixed glycerine with a mixture of the strong acids, nitric and sulphuric acids. The resultant product was found to be highly dangerous and sensitive to impact and friction. He had, in fact discovered nitroglycerine.

Nitroglycerine is a straw coloured oily liquid which, because of its sensitivity and unpredictability remained a chemical laboratory curiosity until examined by the Swedish Military Engineer and chemist ALFRED NOBEL

Nobel was familiar with the military explosives of the day, mainly based on Blackpowder. He was struck with the idea of using Sobrero's discovery for commercial blasting and began to manufacture Nitroglycerine or Blasting Oil on a production basis. The extreme sensitivity of Nitroglycerine to impact proved to be too great a hazard and he set about finding a way of transporting it safely. After many experiments with different absorbing materials, he finally tried kieselguhr in 1866.

Kieselguhr consists of the skeletons of the minute sea creature the DIATOM and has the ability to absorb many times it's own weight of oil.

Kieselguhr had been known for centuries. Nobel found that kieselguhr was capable of absorbing nine times it's own weight of liquid Nitroglycerine giving a red powder which he found was safe to handle. It could be struck with light blows by a hammer and even burned in small quantities without detonating.

It required a minor detonation from the military percussion cap, mercury fulminate to cause detonation. When initiated in this way, however, the mixture detonated with all the violence of the liquid nitroglycerine. Nobel called this mixture after the Greek word for power dynos and the base diatomite, that is :

DYNAMITE.

Prepared as a powder, the Dynamite was made up into paper cartridges for convenient use.

This major invention was completed by devising a safe method of initiating the Dynamite, the combination of Safety Fuse and the Detonator.

The Detonator Nobel designed was based on an earlier discovery (c 1800) of mercury fulminate, which on ignition explodes with little gas but a violent shock. Nobel put a little of this (about a gramme) in a small copper tube. The mercury fulminate is sensitive to flame and is detonated by the spit of flame from the end of the Safety Fuse. A length of Safety Fuse is inserted into the detonator. The detonator is firmly "crimped" onto the fuse. A hole is then made in the Dynamite cartridge with a wooden or brass dowel. This "primed" cartridge is then placed into the shot-hole and the fuse is lit. The shot firer has one second for every centimeter of fuse to get to a place of safety. This three part invention gave to the world the

HIGH EXPLOSIVE

and revolutionised mining, civil engineering and, inevitably, military propellants and engineering.

This fuse fired plain detonator was soon refined to make use of electrical firing using an electrically initiated matchhead instead of safety fuse and the Safety Explosive System was complete.

Alfred Nobel was a remarkable man. Much of his childhood was unsettled and poverty stricken but he became an inventor imbued with a surprising talent and driving force. Unlike most inventors he combined technical creativity with commercial flair, both to a very high degree. He had no capital but that was never a hindrance to Nobel. He used his powers of persuasion to convince the hard-nosed Paris firm of Credit Mobilier to loan him the money which within ten years was to found an international group of Dynamite Companies.

The thirty years between 1870 and 1900 were ones of intense activity during which virtually the whole range of explosives in modern use were devised and the industry expanded on an international scale. The financial rewards for this scientific development were, in the context of the time, beyond any hitherto achieved as evidenced by the setting up of the internationally respected Nobel Foundation and the annual Nobel prizes.

In Great Britain the first charge of Nitroglycerine was produced by the British Dynamite

Company on the west coast of Scotland on the 13th January 1873 and within two years the Company had cleared its capital and established an enviable turnover. The Company continued to prosper and under Alfred Lord McGowan who joined the company in 1894 eventually merged with Brunner Mond, united Alkali and British Dyestuffs and led to the formation of I.C.I. in 1926.

Today there is a bewildering array of explosives and devices available to the civil and mining engineers. The efficiency and effectiveness of the modern high explosive depends on the fact that it detonates irrespective of the conditions under which it is fired, even in the open. It's effectiveness is governed by two important factors. Firstly the power, that is the ability to do work which, in turn, is dependent on the large quantity of gases produced from the small quantity of solid explosive. Secondly the velocity of detonation, that is the speed with which the reaction occurs and which dictates the shattering effect or brisance of the explosive.

The ability of the high explosive to operate on solid material is however primarily dependent on the supersonic shock wave that is invariably produced when the explosive detonates.

The secret of success in using modern explosives in blasting rock in a quarry is to bounce this shockwave off the free face of the quarry. The shock wave, in returning back to the shothole, produces a network of fissures by tensile fracture in the rock so that the enormous gas pressure contained within the shothole can then burst the fragmented rock free from the rock face and blast it onto the quarry floor. The explosive thus operates by pulling the rock apart before blowing it free.

This principle of operating to a free face and breaking the rock by tensile fracture caused by the reflected wave is of major importance in quarry blasting where the face is fired in a series of rows of shotholes fired in sequence using delay detonators or cordtex and detonating relays. The rock is broken out rather like cutting slices of cheese. Each successive line of shots creating a new free face to which the next round of shots can operate.

The principle is of even greater importance in

tunnel blasting particularly where fragmentation of the rock is of importance. The round is in three parts. The first part is the CUT which is fired first to create the FREE FACE to which the subsequent shots operate. The second part of the round is the EASERS which expand the cavity creating a bigger FREE FACE. The final part of the round is the TRIMMERS which are fired last and expand the tunnel to its final cross-section. The use of explosives in underground tunnelling is not, therefore the mere application of brute energy but the careful application of scientific principles in a disciplined and controlled manner.

The introduction of Dynamite and the High Explosive in the 1890s revolutionised mining and civil engineering practice and mining engineering began, from that time, to make dramatic progress. In the very early days the new Dynamite was substituted for the old Blackpowder charges almost on a one for one basis.

Mining techniques soon improved however and, along with that improvement, specialist machines to drill the shot holes and handle the now vastly improved rock yields, were introduced.

Today underground mining is a complex and sophisticated procedure in which high engineering technology is combined with the almost unbelievable energy that is contained in the modern high explosive.

The application of a high explosive charge in a quarry or civil engineering operation is, of course, one of the simplest and most widely used applications. In the sophistication of today's technology, the chemically produced shock wave system of the high explosive finds many more elaborate engineering applications in both industry and in engineering.

This shock wave system can be manipulated in much the same way as any other wave system, for example, light or sound. In the manipulation of light waves the use of lenses can, by different refractive indices, produce either converging or diverging light beams. A similar manipulation can be carried out with explosive stress waves using, in place of transparent media of different refractive indices, explosive media of different detonating velocities. In this way explosive shock waves can be concentrated to produce penetrating jets.

The detonation wave can be concentrated into a jet in which the whole (or most) of the energy of the explosive is concentrated into an intense and powerful penetrating jet focused in one direction. This "Jetting" is accomplished by means of employing a malleable metal cone to form the jet and embedding this cone in a high velocity explosive.

Today this type of sophisticated use of explosives energy has been extended into a bewildering variety of engineering applications only a few of which can be mentioned as illustrations,

The furnace tapper
The underwater cutting stook
Perforation charges for oil wells
Separation charges on space vehicles

Explosives have also made a major contribution in the search for that other vital power source oil and gas.

In the exploration phase for oil the technique of siesmology is used to prepare a three dimensional map of the earth's crust down to many thousands of meters on which map the expensive drilling and exploration programme is based. This technique uses a beam of sound to penetrate the earth and reflect back acoustic signals from the various layers of different rock comprising the earth's mantle. The principle is the same as that of the ASDIC or SONAR systems for detecting underwater objects such as submarines.The principle difference between the SONAR system and that required for oil exploration is that a much more intense beam of sound is needed to penetrate the earth's crust to depths of 8000 meters and still be capable of being reflected back to the surface to provide a readable signal that can be picked up by sensitive Geophones. It, in fact, requires a miniature earthquake and there is no better way of providing that man-made earthquake than by using explosives.

The system is particularly suited to off-shore exploration and played a very large part in the successful search for oil in the Gulf of Mexico, the North Sea and in many other places. The system fires a 25 kilo charge of explosive every 30 seconds from a shooting boat as it steams at 8 knots. This is equivalent to firing a charge every 1/8 of a nautical mile.The recording boat tows a cable in which Geophones are spaced at intervals of 1/8 of a nautical mile so that each reflection is recorded as many times as there are Geophones in the cable. This multi-recording enables modern computing techniques to add (stack) the individual signals together to give clear signals and definitive maps from which the precise location of possible oil or gas wells can be deduced.

The use of explosives in the oil industry does not, however, stop here. Apart from the dramatic use of explosive to extinguish oil well fires, a technique made famous by Red Adair of cinematographic fame, they are being used to solve major engineering problems associated with the production phase. Once the well has been drilled it is "capped" on the sea bed and the oil drawn off by undersea pipeline. In deep waters laying such a pipeline from the surface causes damaging stresses in the pipe which are dangerous and could cause fractures in the pipe with resultant spillage and pollution. One method of eliminating this danger is to weld the pipe on the sea bed as it is being laid.

This is now possible using explosive welding techniques in which a remotely controlled submersible device brings sections of the pipe together on the sea bed and then automatically wraps a thin layer of sheet explosive round the junction. The explosive is fired remotely from the control ship and the high temperature and pressure of the shock wave effectively welds the two sections together without causing any bending stresses

In retrospect the increasing use of explosives in underwater engineering where there is no oxygen is logical since explosives contain all the oxygen they need for complete reaction in their ingredients and are ,therefore, just as effective in producing energy underwater as in air. Explosives are, therefore, ideally suited to provide high energy in airless conditions. It is for that reason that, today, it is the explosive that is taking man across his final frontier - Space.

The principle use of explosives in space vehicles is in providing the enormous amount of sustained energy required to lift the space vehicle from the ground and accelerate it to escape velocity. This, today, can only be achieved by the quasi-detonation in the thrust motors of the chemical rocket. These giant motors operate generally on liquid explosive mixtures or, in the case of the smaller guidance systems rockets, on solid propellant explosives.

All except the smallest space vehicles are assembled in three stages which are discarded in succession to lighten the payload as the vehicle accelerates. The separation of these stages from the rest is accomplished by firing a shaped cutting charge in the form of a ring round the vehicle which is fired either remotely from the ground control centre or manually from the command module and cuts the used rocket stage away from the next stage. The next stage is then fired up using a remotely operated firing system based on and similar to the detonator. In other words the whole system uses explosives in one form or another in all its stages.

The dependence on explosives does not, however, stop once the space vehicle has reached its cruising trajectory or orbit. There are numerous other in-flight and landing operations to perform all of which are carried out by explosive devices or gas operated micro-switches. The use of this form of energy is again logical as the devices are mini power houses that are ideally suited to remote radio controlled operation and can deliver the necessary energy for minimum payload while operating in an airless environment.

It is a sobering thought that, as we enter the Space Age, progress is dependent on the energy producing chemical reaction that first lifted man from the darkness and ignorance of his Neanderthal beginnings and enabled him to attain the enlightenment which led to civilisation and intellectual dominance of this planet. The chemical energy of the combustion reaction has served the human race throughout the millions of years of its existence, has sustained its survival and, in the sophistication of the Explosive, has made possible engineering feats which would have been regarded as miraculous 150 years ago.

Today man is more knowledgeable about the galaxy in which our planet earth is located than ever before. There is ,however, cause for philosophical reflection in the fact that as we approach the twentyfirst century it is a million year old discovery that is today enabling a new generation of scientists to prepare to leave this planet and seek a destiny in the stars.

Chemistry and the Development of the Nuclear Fuel Cycle

C.B. Amphlett

(FORMERLY UKAEA HARWELL LABORATORY)
FARM FIELD, WOOTON VILLAGE, BOARS HILL, OXFORD OX1 5HP, UK

1 INTRODUCTION

In 1938 Hahn and Strassmann demonstrated by classical chemical means that the elements formed when uranium is bombarded with thermal neutrons were of lower atomic number and not, as Fermi had postulated in 1934, transuranic elements. It was shown that this process of nuclear fission was a property of the isotope U235, present to the extent of 0.7% in natural uranium; considerable energy is released, and 2-3 neutrons are emitted for each neutron absorbed, so that in principle a self-sustaining chain reaction capable of releasing large amounts of energy is possible. The potential consequences were soon recognized. In August, 1939, Einstein and Szilard wrote to President Roosevelt about the possibility of constructing a fission bomb, while in March, 1940, Peierls and Frisch alerted the British Cabinet to the possibility. In America the National Academy of Sciences was asked to comment on the Einstein-Szilard note, and in Britain the MAUD committee was set up in April, 1940, to report on the Peierls-Frisch memorandum. Meanwhile McMillan and Abelson had shown in 1939 that the predominant uranium isotope, U238, absorbs slow neutrons to give U239, which is transformed by successive β-decay processes into a fissile isotope of plutonium, Pu239.

In July, 1941, the MAUD committee produced two reports entitled "The Use of Uranium for a Bomb" and "Use of Uranium as a Source of Power"; in October the British effort was consolidated in the Tube Alloys project, responsible to DSIR. In December, 1941, the Manhattan Project was set up in the USA, involving universities, industry, and ultimately a number of new major laboratories. The principal centre for chemical R and D was the so-called Metallurgical Laboratory at the University of Chicago, which was the forerunner of the Argonne National Laboratory. In Britain, work began in several university departments and at

ICI General Chemicals in Widnes in the summer of 1940. Chemical research centred principally on the Chemistry Department at the University of Birmingham and the Clarendon Laboratory at Oxford, with close contacts with ICI. After the MAUD reports were published, the members of the university and industrial teams continued to work for the Tube Alloys project.

2 THE WARTIME PROJECTS, 1940-1947

Two approaches were available for the production of fissile material. U235 could be separated from U238 - gaseous diffusion through a suitable membrane was the favoured route - or alternatively Pu239 could be separated chemically from irradiated uranium. Both routes were followed in the USA, but initially the British concentrated on the former, since knowledge of plutonium chemistry was non-existent; unlike the Americans, they had no access to cyclotrons which could be used to produce another plutonium isotope, Pu238, which could be used to study plutonium chemistry and devise a process for separating Pu239 from irradiated uranium. At that time, the reactors in which uranium would be irradiated were not even designed.

Initially the British chemists concentrated on methods of producing pure uranium metal and pure uranium hexafluoride, UF_6, which was suggested for the gaseous diffusion process because of its volatility. Although it is highly hygroscopic, forming HF and non-volatile UO_2F_2 in contact with traces of moisture, and also a very effective fluorinating agent, attacking most metals and organic compounds, its physical properties are highly suitable, e.g. its vapour pressure at room temperature is around 80mm and it sublimes at 56°C under 1atm. pressure. Extensive studies of the compatibility of materials towards UF_6 showed that it could be safely contained in metal plant which had been adequately prefluorinated, while the purity of samples of UF_6 prepared by ICI improved impressively as experience was gained in its handling and production. Samples arriving at Birmingham in 1940 sealed in Pyrex ampoules were pale yellow powders which might contain only a few percent of UF_6 itself, the residue being mainly UO_2F_2 together with HF and SiF_4; by 1942, water-white crystals were produced which sublimed leaving hardly any residue. Larger quantities were stored and transported in metal cylinders fitted with valves and UF_6-resistant gaskets. A range of lubricants and elastomers resistant to UF_6 was produced by fluorination of organic compounds with reagents such as F_2, $C\ell F_3$ and CoF_3, and a new science of fluorocarbon chemistry was developed. Considerable experience in handling highly-reactive materials such as F_2, $C\ell F_3$ and other interhalogen compounds arose as a by-product of this work. The search for alternatives to UF_6 in America and in Britain led to the synthesis of compounds such as uranium pentaethoxide and uranium tetraborohydride, but none was suitable; the success of the chemists and engineers in handling UF_6 is shown by the fact that all isotope separation plants to date - both gaseous diffusion and gas centrifuge - use it as a working fluid.

The responsibility for production of uranium metal and UF_6 was given to ICI General Chemicals via a contract for £5,000 (more than £100,000 in today's values) to produce 3kg UF_6 by direct fluorination of U metal. This work was later expanded, and in 1943 ICI were producing 50-100kg/day of UF_6 and 450kg/week of cast uranium metal. In the summer of 1940 the chemists at Birmingham produced small quantities of pure uranium metal for ICI to use in developing the production of UF_6. The method involved the electrolysis of molten sodium uranium fluoride to deposit uranium on a molybdenum anode; the melt was very sensitive to traces of moisture, and the finely-divided product was highly pyrophoric, sometimes leading to spectacular pyrotechnic displays. In Spring, 1941, ICI took over metal production at Widnes by a less dramatic route involving the reduction of UF_4 with metallic calcium, which required additional plants for calcium distillation and uranium recovery by remelting. This plant was later expanded threefold. ICI were also given a contract to examine the chemical engineering problems of a gaseous diffusion plant; membrane development began at Kynochs (now Imperial Metal Industries), supported by work at the Clarendon Laboratory in Oxford. By 1943 work was under way in Britain on methods of producing uranium metal and UF_6, and on the use of the latter in a diffusion process; meanwhile our knowledge of the chemistry of uranium and its compounds had expanded considerably.

In 1943 most of the British effort moved to Canada, where it was planned to press ahead with the second route based on plutonium. Between 1943 and 1946 a joint British/Canadian project was set up, first in Montreal and later at Chalk River; the team included a number of French scientists, some of whom had been working in Britain since 1940, who in 1945 formed the nucleus of the French project. A considerable degree of collaboration already existed between the British and American projects, and several British scientists were working at American sites, but in some areas official collaboration was slight or even non-existent. These included plutonium chemistry, separation processes, and the production technology of UF_6 and uranium metal. There was therefore much to occupy the chemists in Canada. Fortunately, official decisions were slow in communicating themselves to individual US laboratories, and in 1944 one of the French scientists, Bertrand Goldschmidt, in the course of a private visit to Seaborg's laboratory in Chicago, obtained not only information on American activities, but, more importantly, 3µg of plutonium and samples of fission products, thus enabling work on plutonium separation to begin in Montreal. The American method of separating plutonium from uranium and fission products was known to be batch coprecipitation with bismuth phosphate, but the Montreal team preferred the alternative of solvent extraction; ether extraction is used in concentrating uranium from pitchblende, and it was expected that plutonium would similarly be extracted. In June, 1944, the Montreal team began by screening 250 solvents, and by October, 1945 they had produced a flowsheet for extracting plutonium by a batch process, using the solvent triglycoldichloride, $C\ell CH_2CH_2OCH_2CH_2OCH_2CH_2C\ell$.

In 1945 Britain and France decided to set up atomic energy projects, in which the development of a method for reprocessing irradiated fuel was an essential task. The British decided to develop a continuous solvent extraction process which would recover both uranium and plutonium. The British team at Chalk River, led by R. Spence, first surveyed the best solvents from Goldschmidt's original list of 250; they chose Butex (dibutoxydiethyl ether) on the grounds of stability to heat and irradiation and high decontamination factors for uranium and plutonium with respect to fission product activity. If the extraction is carried out in nitric acid solution, no additional salting-out agent is needed to drive the extraction of U and Pu into the organic phase; consequently the aqueous raffinate containing the highly-active fission products can be concentrated by evaporation, so reducing the volume of active waste to be stored. In contrast, the Americans were working on a process using hexone (methyl isobutyl ketone) as a solvent, which requires a salting-out agent such as $A\ell(NO_3)_3$, giving active wastes which cannot be concentrated without depositing sludges.

Spence's team worked with 20mg of plutonium which had been extracted from four irradiated fuel pellets supplied by the Americans; this had to be recovered after each experiment for further use, the experiments being designed to reproduce individual extraction stages in a column by a series of batch equilibrations. Column experiments were carried out on uranium extraction, and plutonium was assumed to behave similarly. Work began late in 1946, and in autumn, 1947, the feasibility of a counter-current solvent extraction process had been established, although further work was recommended at higher Pu concentrations and higher fission-product activities. The suggested process used five packed extraction columns. Fuel was dissolved in nitric acid and conditioned so that uranium and plutonium were present as U(VI) and Pu(IV) respectively. These were extracted into Butex in column 1, the highly-active fission products remaining in the aqueous phase. In column 2, plutonium was reduced to Pu(III) and extracted into the aqueous phase. Column 3 was used to salt out uranium into the aqueous phase, and plutonium was further purified in columns 4 and 5. The final products were uranium, plutonium and fission products in aqueous solution, decontamination from fission products being sufficient to enable uranium and plutonium to be handled safely. With the process defined, most of the British team left Canada in 1947 to join the project at Harwell.

3 THE POST-WAR PROCESSES TO 1960: CIVIL APPLICATIONS

At the end of the war the Americans had plants in full production for the production and enrichment of uranium, and for the separation of plutonium produced in their reactors at Hanford. In Britain the decision to set up a uranium production plant at Springfields was announced in March, 1946, and in July, 1947, the reprocessing site to be built at Windscale (now Sellafield) was announced, with a target date in early 1951 for the first active run. Under the conditions in post-war Britain this was a punishing schedule. The work required to flesh out the process recommended by Spence was initially shared between Harwell Chemistry Division, the remaining British scientists at Chalk River, and ICI General Chemicals; later the newly-constituted Production Group under Christopher Hinton took up the work. Harwell worked on refining partition data, the chemistry of solvent extraction, Pu chemistry and purification, effluent treatment and a number of longer-term aspects of the chemistry of U, Pu and individual fission products. ICI compiled flowsheets and constructed a small inactive pilot plant at Widnes, which was later transferred to Harwell; they also supervised the construction and operation of inactive columns at Springfields. After the first plutonium had been produced at Chalk River in 1948, the results of Spence's work were confirmed at higher Pu concentrations in an active pilot plant at Chalk River. Despite several problems the first separation plant was completed at Windscale in April, 1951, and the first active run was completed in June, 1951, several months before the American solvent extraction process began operating. This plant met all specifications and operated satisfactorily for 12 years until a larger plant was needed for the civil programme. Ancillary plants for plutonium production were equally successful, the first billet of Pu being cast at Windscale in March, 1952. Initially, Butex was obtained from America, but production was later set up in Britain at British Industrial Solvents, with ICI producing the intermediate Chlorex.

Steps were also taken to produce uranium to fuel the reactors and for production of UF_6 for enrichment. By 1947 ICI was producing 1400kg/wk of cast uranium, starting from UO_2 supplied by the Americans; when these supplies were discontinued, they were asked to develop a process for the production of cast uranium at Springfields, starting from crude ore. After dissolving crushed pitchblende in acid, radium and its daughters were filtered and crude UO_4 precipitated with hydrogen peroxide. This was dissolved in HNO_3 and evaporated, and uranium was extracted with ether; after back-extraction into aqueous solution, ammonia was added to precipitate ammonium diuranate, which was ignited to UO_3, reduced in H_2 to UO_2 and converted to UF_4 by treatment with gaseous HF. The final stage was to reduce UF_4 to U with metallic calcium, the CaF_2 slag floating on molten uranium. Once again the timescale did not allow the luxury of a pilot plant before plant design could commence. For this reason, and because of the novel processes and potential hazards involved, a batch process was chosen; large-scale production of H_2O_2 and HF was developed, with Imperial Smelting Corporation providing invaluable help in the latter case. Metallic calcium was initially imported from Canada, but was later replaced by cheaper, readily available magnesium. In October, 1948, the plant completed its first full production run to produce cast uranium from Congo pitchblende.

The plants and reactors operating in the 1950s were designed to produce plutonium for military purposes, but the growing interest in nuclear generation of electricity, stimulated by the First Geneva Conference on the Peaceful Uses of Atomic Energy in 1955, led to an expansion of programmes and to changes in the technology. Although the basic principles are the same, the requirements of military and civil programmes are very different. Plutonium for military use should be predominantly Pu239, since the higher isotopes Pu240 and Pu241 produced by successive neutron capture have undesirable nuclear properties. Consequently the degree of burnup of uranium fuel is limited before reprocessing takes place. In a civil power reactor, too frequent reprocessing would be uneconomic, and so the fuel is irradiated for longer periods. The theoretical maximum limit in a thermal reactor fuelled with natural U is set by the U235 content of the fuel, viz 0.7%; even this is not attainable in practice, because of the production of neutron-absorbing fission product poisons which further decrease the reactivity in addition to the decrease resulting from the consumption of U235. Furthermore, the need to operate at higher fuel temperatures in order to increase the thermodynamic efficiency of the system requires a fuel which is more resistant to thermal and irradiation damage. The natural consequence of the move towards higher temperatures and longer irradiation times before reprocessing was the development of ceramic fuels such as uranium oxide, enriched with U235 to about 2-5%. Uranium which has been reprocessed can be enriched again by blending with material from an enrichment plant, increasing the power produced for a given mass up to tenfold. The next stage of development is to tap the potential of the unused U238 by fission with fast neutrons in a fast reactor, with the aim of 10-20% burnup in a single pass, which may be increased further by recycling. Thermal reactors operating on slightly-enriched uranium would produce power and plutonium; after reprocessing to recover U and Pu, the former could be enriched for further use, either with U235 or Pu. The latter can also be used in mixed oxide fuel (PuO_2:UO_2)

in the fissile core of a fast reactor, with natural or even depleted U in a breeder zone to produce more Pu than has been consumed. The large stocks of depleted uranium would be utilised to realise the full potential of uranium.

The economic competitiveness of nuclear power depends on many factors which have changed frequently and cyclically over the past 30 years. They include estimates of energy requirements, the availability and cost of alternative fuels such as coal, gas and oil, availability and cost of renewable energy sources, the estimated reserves and cost of uranium, fiscal policies, as well as technical developments and environmental considerations (as regards both nuclear power and conventional sources). Although there is considerable debate about the desirability and competitiveness of nuclear power, there is little doubt that continued development is essential in the long term, particularly when we consider the growing energy demands of less-developed countries.

4 DEVELOPMENTS IN PROCESSES SINCE 1960

In meeting the challenge to produce nuclear power competitively, chemistry and chemical engineering skills have made notable contributions in three key areas:

(i) New fuels and simpler production methods
(ii) Improvements in reprocessing
(iii) Reduction of environmental impact by developments in waste management.

Expanding post-war programmes led to a search for sources of uranium to supplement the rich pitchblende deposits, and many new, lower-grade deposits were discovered in Canada, America, South Africa, Australia, France and Russia. New methods were developed to upgrade the uranium concentration at source before shipping the product to the users. In South Africa sulphuric acid leaching was used to produce anionic sulphate complexes which were absorbed on a strongly-basic anion exchanger, while in Colorado and Canada acid leachates were extracted into trialkylamine solvents diluted with kerosene, the amine acting as a liquid anion exchanger. In both processes, uranium was subsequently extracted into aqueous solution and precipitated with ammonia. The fuel production plants now operated with ammonium diuranate as starting material, and ether was replaced by less flammable tri-n-butyl phosphate in the uranium purification stage. The first major change was the replacement of batch processing with continuous, fluidized bed denitration to UO_3, followed by reduction with H_2 to UO_2, and hydrofluorination with gaseous HF to form UF_4. Careful control of the denitration step is essential, since the efficiency of reduction and hydrofluorination depends critically on the porosity and crystallite structure of UO_3

formed in the first stage. Production of fuel at Springfields followed this route until 1971, when an entirely new concept was introduced, the Integrated Dry Route (IDR) for production of oxide fuel for Advanced Gas-cooled Reactors (AGR). This uses UF_6 as feed material, converting it directly to UO_2 powder by hydrolysis and reduction with a mixture of steam and hydrogen in a single-stage, electrically-heated kiln. Either non-irradiated UF_6 or enriched UF_6 from reprocessed material can be used as feed, and the process allows complete flexibility in regard to the source of feed and the U235 content of the product. The original plant had a design output of 250te/yr of product, and has processed more than 12,000te since 1971; the process has been licensed in France and in America. A larger plant is planned for the mid-1990s, to produce 230te/yr as AGR fuel, 70te/yr as PWR fuel for domestic programmes, and 200te/yr UO_2 for export, with potential to expand PWR fuel production to 200te/yr. Future fuel development programmes include the production of mixed oxide fuels (MOX) containing UO_2 and PuO_2, either for thermal or fast reactors. Research and development programmes are under way in Belgium, France and Britain; France is committed to plutonium recycling in PWRs, and there are plans for MOX fabrication plants in France and Britain.

Significant changes have also been made to reprocessing methods, and chemistry has again played a major role. The first plant at Windscale (1952) used Butex to separate uranium and plutonium from fission products, and tri-n-butyl phosphate diluted with odourless kerosene (TBP/OK) to separate plutonium from uranium after reducing the former to Pu(III) with ferrous sulphamate. The first American plant at Hanford used the Redox process, with hexone as solvent and $Al(NO_3)_3$ as salting-out agent; this was later replaced by the Purex process at Savannah River, using 20% TBP/OK and no salting-out agent. The second British plant began to reprocess civil reactor Magnox fuel at Windscale in 1964; this used TBP/OK throughout, because it gave a better decontamination from fission products in the first separation stage and in the medium-active wastes. Other variants included the use of trilaurylamine to separate U from Pu in the first French plant and the use of ion exchange to purify and concentrate plutonium in later stages. The latest plants, using TBP/OK, are designed to reprocess oxide fuel from civil reactors. In France a series of plants is being built at Cap la Hague, with throughput increasing to 1600te/y in the 1990s. The British equivalent THORP (Thermal Oxide Reprocessing Plant) is planned for the mid-1990s with a capacity of ca 1800te/yr, compared to a figure of 1500te/yr for the existing civil plant. The French plants are expected to be used for both UO_2 and MOX fuels from 1997. No comparable developments in reprocessing plant design have taken place in America, where reprocessing is temporarily in abeyance.

The separation of uranium and plutonium from fission products in the first stage of a solvent extraction process involves the extraction of uncharged or anionic nitrato complexes of U(VI) and Pu(IV), the mainly 2- and 3-valent fission products remaining in the aqueous phase. Ruthenium does not behave ideally, and can lead to undesirable contamination of the products. Extensive chemical studies showed that it forms very stable complexes of the ion $RuNO^{3+}$, e.g. $RuNO(NO_3)_3aq_2$ and $RuNO(NO_3)_2OHaq_2$, which dissolve in organic solvents and are only hydrolysed slowly to non- extractable species. Very careful control of conditions after dissolution of the fuel is necessary to minimise extraction of ruthenium in the first stage. Other elements which are extracted to some extent include technetium and neptunium; since higher burnup leads to higher concentrations of these, additional redox/solvent extraction processes are required to remove them in the U and Pu purification stages. The chemical consequences of introducing new processes are well illustrated by the effects of replacing ferrous sulphamate by U(IV) to reduce Pu(IV) to Pu(III) before separation of U and Pu. By making this change no new elements are added to the system, and the resulting salt-free medium active wastes can be combined with the highly-active wastes for vitrification. However, in order to stabilise the system against re-oxidation, hydrazine is also added, and it was found that technetium, which is co-extracted with U and Pu as a complex of pertechnetate with fission-product zirconium, efficiently catalyses the destruction of hydrazine by nitric acid. Attempts to minimise Tc extraction by reducing pertechnetate to non-extractable forms or by preferential complexing of Zr unfortunately involved reagents which also react with U and Pu, and the solution finally adopted was to change the mode of operation of the extraction column in order to minimise the time available for the catalysed decomposition of hydrazine to take place.

Although there is much debate about the need for fuel reprocessing, it seems more sensible to recover unused uranium and plutonium for recycling, and to store fission-product wastes safely in highly unleachable matrices, rather than leave them in stored fuel elements which are less resistant to breakdown and corrosion over very long periods. This approach is being followed in Western Europe, where BNFL and Cogema are reprocessing fuel from Britain, France, Germany, Japan and other countries at Sellafield and Cap la Hague. More than 30,000te fuel have been reprocessed at Sellafield alone since 1952, and 25,000te U are expected to be recovered from the reprocessing of spent oxide fuel during the 1990s. The ability to recycle fissile material will become more important when problems of uranium supply arise in the next century. As a measure of the scale of these operations we may note that the BNFL Annual Report for 1988/89 indicates a turnover of £916M and operating profits of £275M for fuel production, enrichment and reprocessing; exports totalled £169M.

Considerable advances have been made in radioactive waste management, driven by reduced levels of permitted discharges, greater environmental pressures and technical advances in operating practice. Discharges to the environment are reduced as far as is reasonably achievable within current discharge limits by appropriate chemical treatment, and all other wastes are converted into immobilised solid forms which can be safely stored in carefully-engineered repositories. A major source of low-level waste activity is the water from the cooling ponds in which fuel elements are stored before reprocessing; corrosion of Magnox cans leads to release of soluble, long-lived Cs137 and Sr90. Their contributions to the activity discharged may be reduced by factors of 200 and 500 respectively by adjusting the pH with CO_2 and removing them by ion exchange on the zeolite clinoptilolite. A plant was commissioned for this purpose at Sellafield in 1985 (SIXEP, Site Ion Exchange Effluent Plant) after work at Sellafield and Harwell. Activity levels in effluents arising mid-way through reprocessing, which contain actinides and certain fission products, have been reduced by evaporation prior to treatment with highly-active wastes. A new treatment plant, EARP (Enhanced Actinide Recovery Plant), to be commissioned at Sellafield in 1992, will reduce discharges still further by co-precipitation and absorption on a ferric oxide floc; the latter will be dewatered by crossflow ultrafiltration on zirconia-coated graphite tubes manufactured by the French company Techsep, to give a settled floc with up to 15% w/w solids and a crystal-clear permeate, with very efficient removal of actinides. Good removals of Ru and Sr can be achieved by adjusting the pH, and Cs can be removed by adding the complex cyanide $Na_2NiFe(CN)_6$, which exchanges sodium for caesium.

For many years highly-active wastes have been stored after evaporation, either as concentrated acid solutions in Europe, or as saturated solutions, salt-cake or sludges in America. Several processes have been investigated for converting them into solid forms, e.g. pot or fluidised bed calcination to oxides (USA), ion exchange on granulated montmorillonite which was then fired to fix the activity (Britain), and incorporation into mineral matrices such as nepheline syenite (Canada) and the synthetic material Synroc (Australia). A parallel approach involving incorporation into glass was pioneered in Europe; an extensive programme on glass composition, uptake of waste, resistance to ground-water leaching and irradiation damage, culminated in the development of the French AVM process, first operated at Marcoule in 1978. The ideal glass composition, matched to the waste, is 35-50% SiO_2 and 14-20% B_2O_3. The waste solution is evaporated to dryness and roasted in a rotary kiln; the product is fed to a container in a furnace, mixed with granular frit and melted at 1150°C, after which the container is sealed. Off-gases are filtered, scrubbed and recycled. The Marcoule plant has a throughput of 200 canisters per year, each containing 150ℓ of glass. By 1987 all high level wastes stored in solution in France since 1968 had been converted into glass. Two similar plants have been built at the Cap la Hague reprocessing facility, each with two lines operating and one on standby; each plant will process 600 cans a year, each containing 400kg of glass. The process has been licensed to BNFL, which will operate two plants at Sellafield to vitrify the backlog of $1300m^3$ of waste solution stored over 30 years, as well as future arisings. Each container will contain glass equivalent to the waste from 8te of Magnox fuel or 2te of PWR fuel. Low and intermediate level wastes will also be converted to solid form in chemically stable matrices such as cement or bitumen.

After preliminary cooling in surface stores, vitrified high-level wastes will be stored in deep repositories. The canisters will be surrounded by material which can absorb any activity which escapes, e.g. bentonite or clay, and the repository will be situated in a geological medium in which water flow is minimal or zero. Several European countries are investigating potential sites, and Sweden has licensed a store in granite below the sea-bed. Over very long periods of time it is expected that corrosion of the containers will allow access of groundwater to the glass, leading to leaching of soluble activity. The barriers to transport include the solubility of key nuclides, saturation levels in groundwater, absorption on the back-fill and concrete walls of the repository, comprising the "near field" effects, followed by the rate of movement of water through the external strata and the retention of activity within the latter by precipitation, adsorption and ion exchange (the "far field" effects). Chemical studies show that near-field leachates are alkaline (as high as pH12.5) and the pH is expected to fall very slowly over millions of years; under these conditions actinide solubility will be exceedingly small. Considerable sorption and ion exchange occur on the concrete and back-fill material. Laboratory and field studies on permeability, porosity and diffusivity show that water flow-rates in suitable strata are very low, while sorption and ion exchange will reduce the rate of activity transport still further. Transport rates may be modified by colloidal effects and by complexing by organic materials present in the wastes or in the ground, e.g. humic and fulvic acids. Laboratory tests on solubility and equilibrium leach testing provide chemical data which can be combined with hydrodynamic modelling of water flow; so far the results suggest that while water from repository depth will reach the surface in 10^4 years, a readily-sorbed nuclide will take 10^6 years to appear. Some confirmation of these extremely long periods is provided by the existence of a number of deep, naturally radioactive mineral deposits where there is no detectable evidence of the activity in surface water. Another striking example of the extremely slow transport of key nuclides is provided by the Oklo deposit in Gabon, where a natural fission reaction took place in an extremely rich uranium deposit some 2×10^9 years ago. Fission occurred intermittently for more than half a million years, temperatures rose to >600°C and the rocks suffered radiation damage. Nevertheless, chemical analysis of the surrounding strata shows that elements such as Np, Pu and Th formed in the chain reaction have moved only very short distances.

REFERENCES

An excellent account of the history of the nuclear projects to 1952 is given in M. Gowing, 'Britain and Atomic Energy, 1939-1945' and 'Britain and Atomic Energy, 1945-1952', MacMillan, London, 1964 and 1974. Subsequent developments are covered extensively in articles in Atom, published monthly by the United Kingdom Atomic Energy Authority.

The Chemistry of Solar Energy

Cleland McVeigh

SCHOOL OF ENGINEERING, GLASGOW COLLEGE, GLASGOW G4 0BA, UK

1 INTRODUCTION

This paper reviews the contribution made by chemists to the science and practice of solar energy, with particular emphasis on chemistry. Starting with a brief summary of the early scientific experiments on the use of "burning mirrors" and the work of Priestley and Lavoisier in the 18th century, developments of the 19th century are traced. These include solar engines, refrigeration, distillation and solar heat collectors. The energy philosophy of Professor Sir Frederick Soddy is then outlined. He was probably the first person to draw the distinction between "capital" and "revenue" energy sources (1921). Modern developments date from the early 1950s. Here the work of the chemist Farrington Daniels is particularly significant, with the publication of his text "Direct use of the sun's energy" in 1964.[1]

The second part of the paper examines trends in current research and development work. This includes photochemical and photoelectrochemical conversion, thin film coatings, chromogenic materials, transparent insulation materials, photovoltaic energy conversion, chemical heat storage and waste destruction. Chemists have already played an important role in this work and the new challenges are waiting.

The Early History

Man has appreciated for thousands of years that life and energy flow from the sun.[2] Socrates (470-399 BC) is thought to have been the first to describe the fundamental principles governing the use of solar energy in buildings.

Many examples of the use of "burning mirrors" or focusing devices are reported in the literature. Authors include the English monk Roger Bacon, at the end of the thirteenth century, Leonardo da Vinci (1515) and the French philosopher Buffon in 1747.[2,3] The British scientist Joseph Priestley used a solar focusing system to heat mercuric oxide in his experiments which resulted in his discovery of oxygen in 1774. This work enabled the famous French chemist, Antoine-Laurent Lavoisier, to develop the correct theory of combustion. A picture of Lavoisier standing on a platform near the focal point of some large mounted glass lenses was also published in 1774.

Experiments to determine the intensity of the sun's radiation - the solar constant - were first carried out at the beginning of the 19th Century by two independent workers, Sir John Herschel and the French scientist Claude Pouillet.[2] Both used the same principle - exposing a known quantity of water to solar radiation and measuring the temperature rise over a given period of time. By 1860 another Frenchman, August Mouchot, had constructed a parabolic mirror which he used to drive a small steam engine.[2] Subsequently he exhibited a "solar pumping-engine" in Paris in 1866 and wrote the first book ever published on solar energy in 1869.[4] A picture of his successful solar-powered refrigerator, producing a block of ice at the 1878 Paris Exposition, is also well-known. Another series of solar engines was also built at that time in the United States by John Ericsson, a Swedish - American, who was also warning about a possible "energy-crisis" in 1876 and predicted that great changes in international relations would occur as the coal fields would eventually become exhausted. The first major development in the use of solar energy for the distillation of water dates from the designs of Charles Wilson in 1872 for a plant at Las Salinas, about 110 km inland from the coast of Chile.[5] This was also the world's largest system for many years, having a total collecting area of some $4750 m^2$ of glass.

Simple solar heat collectors using mirrors and glass covers were also developed for cooking in India by an Englishman, W.Adams, and reported in 1878.[6] One of the earliest solar space heating applications, consisting of a "..surface of blackened slate under glass fixed to the sunny side of a house.." was claimed in 1882, and this was followed by reports on the first use of the flat-plate solar collector, but in an application to a water-pumping system in 1885.[2,7] The first commercial solar water heaters in the world appear to have been manufactured in the United States by Clarence Kemp from Baltimore, Maryland, who patented the Climax Solar-Water Heater a few years later in 1891[8]. The need for water pumping in arid regions where clear sky conditions often give long periods of direct solar radiation led to the most spectacular solar engine development of the time - the Shuman-Boys Sun-Heat Absorber at Meadi in Egypt in 1913. With five tracking parabolic mirror sections each 62.5m long and 4.1m wide, a maximum pumping horsepower of 19.2hp was recorded - a figure which could have been increased to about 100 with a modern low-pressure steam turbine[2].

At this point it can be seen that although only a restricted range of engineering materials was available, the basic principles of many practical applications of solar energy had been understood.[2] However, little work was carried out on solar applications for the next forty years. Power generation, first from coal and later from oil became widespread and relatively inexpensive. It was left to the distinguished chemist Professor Sir Frederick Soddy to draw attention to the problems which are only beginning to be appreciated today. In 1912 he wrote[9]:

> "...this is the beginning of the ages of energy, the Age of the Energy of Coal...unfortunately only too true , and the whole earth is rendered the filthier thereby. Moreover, the age will last just so long as the coal supply lasts, and after that the last state of the race will be worse than the first, unless it has learned better"

His thoughts on the discoveries of new natural resources were also quite remarkable :

> "A find of energy in Nature means an addition to the general wealth, a postponement of the day of bankruptcy, which each new invention of science, on the other hand, brings nearer."

Soddy also spelt out the limitations of economic theories when applied to energy.[10] In two lectures he warned that economists were liable to mistake for laws of nature the laws of human nature and that the principles and ethics of human law and convention must not run counter to those of thermodynamics. He was the first to draw a distinction between the continuous "revenue" of energy received from the sun and the use of the "capital" energy stored in fossil fuels. He predicted that the period of prosperity through which Great Britain had passed was destined to be short-lived because it depended on substituting the capital energy of fuel for labour.

Modern Developments

Modern developments date from the early 1950s. Some work had been carried out by a few dedicated workers in the United States, such as Dr G.C.Abbot. The Cabot Bequest to the Massachusetts Institute of Technology resulted in solar research spreading, both within the United States and to other countries. This resulted in the formation of the Association for Applied Solar Energy (now the International Solar Energy Society) in 1954. One of the driving forces behind the Society at that time was the chemist Farrington Daniels. His book[1], first published in 1964, was subsequently reprinted as a paperback, and is still considered to be one of the best books on solar energy available today. This extract from his preface is still as relevant as it was nearly thirty years ago:

> "Research in the field of solar energy use is unique in several respects. First, it cuts across many different sciences and branches of engineering - physics, chemistry, meteorology, astronomy, chemical engineering, mechanical engineering, and electrical engineering. Often an area between different fields of science that has been neglected becomes a fruitful field of research. Solar energy research is an example."
>
> "Second, it holds promise of leading rather soon to benefits for human welfare. Scientists formerly took little thought of the social and political impact of their work - but all this has been changed since they developed nuclear energy and made atomic warfare possible."
>
> "Third, it can be carried out in small laboratories with inexpensive equipment. Expensive nuclear reactors, atom smashers and wind tunnels are not required, nor is it necessary to master highly specialized techniques....."

Interest in solar energy research and development has grown very rapidly since the early 1970s. By 1976 a survey carried out by the author[2] showed that over forty countries had solar research and development programmes. During the 1980s new programmes were started in many other countries and today the major Solar Conferences will often report work from over 100 countries. The Commission of the European Communities has been particularly active with its collaborative work.[11] This is published on completion and a wide range of reports is available, giving excellent access to current developments and trends. The major Solar Conferences, organized every two years by the

International Solar Energy Society can attract over one thousand abstracts, with socio-economic and educational topics now well-established.

2. THE SOLAR RESOURCE

Radiation is emitted from the sun with an energy distribution fairly similar to that of a "black body" or perfect radiator, at a temperature of 6000K. The value of the solar constant - a term used to define the rate at which solar radiation is received outside the earth's atmosphere at the earth's mean distance from the sun, by a surface perpendicular to the solar beam - is 1.373 kWm^{-2} with a probable error of 1 - 2%[12]. During the year the solar constant can vary by ± 3.4%, partly due to variations in the earth-sun distance. An appreciation of the way in which the levels of solar radiation can vary in different geographical locations is essential for any assessment of a solar application. Figure 1 shows the solar radiation

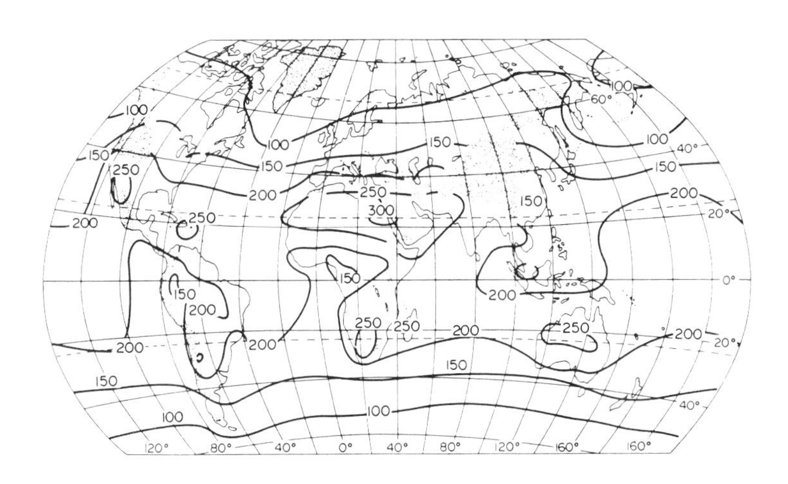

Figure 1. Annual mean global solar radiation on a horizontal plane at the earth's surface, Wm^{-2}, averaged over 24 hours, after Reference 13, with permission.

received on the earth's surface averaged over 24 hours on each day throughout the year. The two major problems are the relatively low maximum intensity - about 1 kWm^{-2} is the maximum within an hour or two of mid-day on clear days - and its intermittency. From Figure 1 it can be seen that the average in the UK is close to 100 Wm^{-2}, while in the Arab countries, such as Egypt and the Sudan it is close to 300 Wm^{-2}. A detailed look at the mean daily totals of solar radiation, averaged for each month in London and Khartoum is shown in Figure 2. In the UK and most of the

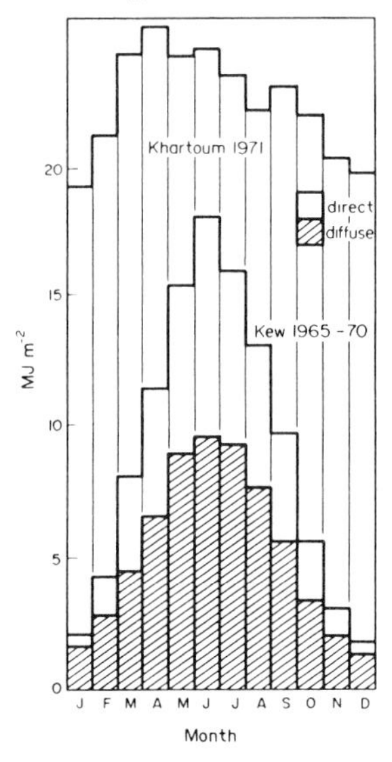

Figure 2. Mean daily radiation on a horizontal plane (after McVeigh[14])

other Northern European countries there is approximately a four-fold variation between the worst three-month winter period and the best three-month summer period. There are two categories used to describe the radiation received on the earth's surface: direct radiation (or direct beam radiation) is not scattered during its path through the atmosphere. Diffuse radiation is scattered by water vapour and other particles in various layers of the atmosphere. About half the solar radiation in Northern Europe is diffuse. In countries with high annual radiation levels there is usually a small variation from one month to the next and there is a much greater proportion of direct radiation. Many countries have greatly increased their efforts to improve their solar data bases since the 1970s with new basic measurements and mathematical modelling.

3 CURRENT RESEARCH AND DEVELOPMENT WORK

Photochemical conversion of solar energy

The direct conversion of solar energy into stored chemical free energy has attracted research workers for many years. For example, Tsubomura[15] cites Ciamician[16] who, in 1912, described his dream of new industry based on solar photochemical conversion which is self-sustaining and pollution-free. An excellent review of the work up to the 1960s by Farrington Daniels[1] included a description of basic processes and suggested that detailed studies of the reaction of photosynthesis (the conversion of carbon dioxide and water into carbohydrate and oxygen) could help to point the way towards research in the quest for new photochemical reactions.

Photochemical processes have been defined by Bolton and Archer[17] as those in which the absorption of solar photons in a molecule produces excited states, or alternatively in a semi-conductor raises electrons from the valence band to the conduction band. As a result of the chemical reactions which may then occur, some of the excitation energy may be stored as chemical energy or a useful chemical reaction may be catalyzed. Tsubomura defines photochemical conversion as a technology to synthetize valuable chemical materials or fuels by the use of solar energy.[15] Approximately half of the total solar radiation which reaches the earth arrives in the uv and visible range (300-700nm) and can be used in various photochemical reactions. This often leads to systems containing highly coloured substances. The other half, which occurs in the infra-red region, cannot make a useful contribution as its energy concentration is too low. The maximum overall efficiency of any photochemical energy conversion is limited to about 30% because some of the higher energy photons of shorter wavelengths have some energy degraded as heat during the reaction. The majority of photochemical reactions are exothermic and are not suitable for converting solar radiation into stored chemical energy. The known endothermic (energy storing) reactions which occur with visible light are, in theory, capable of producing valuable chemical fuels. A major problem has been that most of these endothermic reactions reverse too quickly to store the energy of the absorbed light.

It can be seen that two types of chemical reaction driven by sunlight are outlined above - those in which there is a net storage of solar energy and the product of the reaction could be either a fuel, such as methane or hydrogen, or electricity - and those where the reaction may be photocatalytic and the sunlight would be used to catalyze the synthesis of useful chemicals. Bolton and Archer[17] further divided the subject into two major areas as follows:

(1) Direct processes, in which the direct absorption of solar photons by dye molecules or a semi-conductor leads to useful photochemistry or photoelectrochemistry. Work in this area is normally on a laboratory scale.

(2) Thermal processes, in which concentrated sunlight generates the high temperatures used to drive useful chemical reactions. There are only a few large-scale facilities where high-temperature concentrated sunlight is available.

The general criteria for the selection of useful photochemical conversion and storage systems have been evolving since a 1957 Symposium on photochemistry in the liquid and solid states[18](e.g. Porter and Archer[19], Bolton[20], Bolton and Hall[21],) and have been summarized by Bolton and Archer.[17] However, all the systems investigated up to 1990 were considered by Tsubomura [22] to be unsatisfactory because they failed to meet some of these criteria. He strongly believed that chemical conversion by means of semiconductor photoelectrochemical methods seemed to be more promising. He also pointed out that the technique of electrochemical synthesis combined with solar photoelectric conversion (by the use of solar cells) was more feasible and readily applicable, if not yet economic.

Chemical conversion with photoelectrochemical cells

The distinctive feature of the photoelectrochemical cell is that it can transform solar energy directly into stable chemical energy.[23] Research in this field expanded very rapidly since the work of Fujishima and Honda on water splitting with TiO_2 reported in 1972.[24] The principle of the photoelectrochemical cell (PEC) is that either an n-type or a p-type semi-conductor electrode is immersed in an electrolyte solution together with a counter electrode.[23] With an n-type semi-conductor electrode an electric field is formed inside the semi-conductor which drives the light-induced conductive electrons into the outer circuit and drives the light-induced positive holes out into the electrolyte solutions. Where the solution contains redox agents R/Ox, the reduced species R will transform to the oxidised form Ox. If this is stable it drifts towards the counter electrode where the reverse reaction Ox + e $\rightarrow$ R occurs and no ultimate chemical change has occurred in the solution. As both a photovoltage and a photocurrent are generated in the outer circuit, the PEC is behaving like a photovoltaic cell. Memming[25] lists 14 PECs with different semi-conductors and redox couples, four of which had relatively high (greater than 10%) conversion efficiencies. The advantage of using PECs for producing electrical energy is that they can be fabricated very easily without needing any diffusion or evaporation techniques for making the junction. Their disadvantage is that many of the known semi-conductor electrodes with suitable band gaps are corrosive in aqueous solutions, while those which are stable in water have too wide a band gap.

Applications of thin film coatings

Two important applications of spectrally selective thin film coatings for the efficient conversion of solar energy are with solar absorber surfaces and transparent heat mirrors.[26] The majority of flat plate solar collectors have five main components: a transparent cover, tubes or some form of passage integral with the absorber plate to carry the heated fluid, an absorber plate, insulation and a casing or container.[27] The overall efficiency of the collector system can be improved through the use of both anti-reflection and infrared-reflecting coatings, but a more cost-effective method is to use uncoated glass and treat the absorber surface with a spectrally selective coating having a high solar absorptance and low thermal emittance. Hundreds of different surface coatings have been reported in the literature since their use was first suggested by Tabor[28] in 1956. Transparent heat mirror coatings transmit the incident solar radiation and reflect the longer infrared radiation. The U-value of a double-glazed window can be nearly halved by the inclusion of a suitable heat mirror film. A discussion of recent work in these fields is given by Granqvist[29].

Chromogenic Materials

Chromogenic materials offer the possibility of developing advanced glazings which combine variable control of solar gain with efficient thermal insulation.[30] In a major review of the technical properties and merits of known electrochromic phenomena in 1984, Lampert[31] pointed out that most of the wealth of technical literature and patents dealing with electrochromic materials and devices was primarily for electronic information display or other small-scale applications. Consequently only minor attention had been paid to electrochromic devices as transmissive devices. Since then, transparent apertures employing photochromic, thermochromic or electrochromic materials have been the focus of intensive world-wide research by many groups[32,33] concerned with the efficient use of energy in buildings. The electrochromic window is the most advanced example of these efforts. It is basically a multilayer thin film device which performs as an electric cell, and consists of an electrochromic layer and a counter electrode, or ion storage layer, separated by an ion conductor. For window applications these layers are commonly sandwiched between two transparent electronic conductors which are deposited onto transparent substrates, e.g. glass or polymeric materials. In operation a dc electric field is applied across the transparent conductors and ions are driven either into or out of the electrochromic layer causing reflectance and/or absorptance modulation of visible and near infrared electromagnetic radiation and hence changes in the optical properties of the device. The electrochromic layer may be caused to colour or bleach in a reversible way under the influence of the external electric field. The principal aim of current research is the development of a stable, durable, all solid state electrochromic device - the "smart window".[32]

Liquid-crystal-based chromogenic materials have also been successfully used as electrically activated devices.[34] Two transparent electrodes provide an electric field to change the orientation of liquid crystal molecules interspersed between the electrodes. The orientation of the liquid crystals alters the optical properties of the device. Two main types of liquid crystal systems, the guest-host and polymer-dispersed or encapsulated devices, have been identified for large areas. Their disadvantages are that their unpowered state is diffuse, haze remains in the activated (transparent) state and UV stability is poor.[32] A third approach uses suspended particle devices, but various technical problems such as long-term stability and cyclic durability have slowed their development.

Two non-electrically activated devices use photochromic or thermochromic materials. Their research histories date back at least 100 years.[32] When photochromic materials are exposed to light they change their optical properties, only reverting to their original properties in the dark. Photochromic plastic has been developed for opthalmic use and could become useful for regulating solar glazings.[32] Thermochromic materials display a large optical property change when a particular temperature is exceeded. Above this critical temperature, transmittance is reduced and if this temperature is close to a comfort temperature thermochromism could be used for automatic temperature control in buildings.[34]

Transparent Insulation Materials

A new technology is emerging from European initiatives which will bring about revolutionary changes in the building industry.[35] Transparent (or translucent) insulation materials are a relatively new class of materials which combine the uses of glazing and insulation in the traditional design of any solar thermal system. While the primary use of glazing in buildings has also been to allow light to enter, its ability to transmit radiation gives it the subsidiary function of providing solar heat. Insulation suppresses conduction and convection losses from buildings, but the traditional opaque materials such as polystyrene granules or foam which have been developed for this purpose are equally effective in suppressing solar gain from the outside of the building. For large energy gains both high irradiation levels and high values for the product of the solar transmittance and absorptance of the absorber are essential. The influence of insulation (U-value) depends on the temperature level of the system and the desired heat storage period. For low U-values very good absorption is needed in the thermal wavelengths, although infrared (IR) selective coatings can also be used on the front cover or the absorber plate to reduce IR radiation losses.[36] Convection losses can be greatly reduced by the use of structured materials such as capillaries and honeycombs or low-pressure systems. Until recently no natural or man-made product could offer both high transmission, low conduction, and strong convection suppressant characteristics. But by the mid-1960s it was possible to conceive the potential benefits of such a material.[37] Following rapid advances made in Germany during the 1980s the term "Transparent Insulation" was accepted as best describing the goal of the technology.[38]

The first experimental results were presented in in 1985 by Wittwer et al[39] and the following four generic types which display different physical properties were first proposed in 1986[40]:

Type	Examples
Absorber-parallel	Multiple glazing, plastic films
Absorber-perpendicular	Honeycombs, capillaries
Cavity structure	Duct plates, foam
(Quasi)-homogeneous	Glass fibres, aerogels

By 1990 several systems had become commercially available and considerable potential for more scientific research work had been identified.

Photovoltaic energy conversion

The direct conversion of solar energy into electrical energy has been studied since the end of the nineteenth century.[2] The early work was concerned with thermocouples of various different alloys and efficiencies were very low, usually less than 1%. Modern developments date from 1954 when the Bell Telephone Laboratories discovered that thin slices of silicon, when doped with certain traces of impurities, achieved efficiences some ten times greater than the traditional light sensitive materials used in earlier devices. Since then photovoltaic cells, modules and systems have developed rapidly.[41] Improvements in their cost-effectiveness, reliability and life have resulted in photovoltaics becoming the first choice in a wide range of applications on both engineering and economic grounds. The cost of the photovoltaic module, described as "..the basic building block of the system" has fallen in real terms to about one tenth of its value in 1980 and is expected to continue to fall, unlike almost all other competitive electricity generation systems.[42] Among the advantages listed for the modern solar cell during the 1970s were that it had no moving parts, an indefinitely long life, required little or no maintenance and was non-polluting.[13] The lifetime should now be modified to "probably more than twenty years" and there are the normal environmental hazards associated with the manufacture of semiconductors. Unlike other types of electric generator it is suitable for a very wide range of power applications from a few milli-watts to several thousand megawatts. Photovoltaics is now widely recognised as a mature technology, ready to move into a new phase in an expanded product market which already includes telecommunications, cathodic protection, remote power and utility demonstration projects.[42]

The basic principle of the solar cell is that the doping of a very pure semiconductor with small traces of impurities can modify its electrical properties, producing the p-type, having fixed negative and free positive charges, and the n-type (fixed positive, free negative). If these two types are placed together and the surface is exposed to sunlight, electrons will diffuse through the p-n junction in opposite directions, giving rise to an electric current. Three main types of photovoltaic module design are particularly promising in both performance and cost - the crystalline flat plate, thin film flat plate and concentrator modules. The "standard" efficiency of solar cells reached 37% by 1990, and the physical limits to photovoltaic energy conversion had been established.[41]

Chemical Heat Storage and Waste Destruction

Thermal energy storage is essential for many solar thermal applications. The properties of suitable salt hydrates were first discussed by Telkes in 1974.[43] Sodium sulphate decahydrate, mixed with 3-4% borax as a nucleating agent if complete crystallization is to be obtained, was the most tried material, with a transition temperature close to 30°C. The problem of a barrier being formed between the liquid and solid phases proved very difficult to solve and numerous polymeric stabilisers were tried. Many salt phase change materials have been tried in the past two decades, including calcium chloride hexahydrate and sodium acetate trihydrate, several of which are now commercially available.[44] Recent developments were reported in 1990.[44] Molten nitrate salt receivers have been designed for the 40MWt Solar One electricity generating pilot plant.[45]

Concentrated sunlight is increasingly being used to destroy hazardous chemicals. Another solar application is the photocatalytic detoxification of water.[45]

4 ACKNOWLEDGEMENTS

I am very grateful to Dr Michael Hutchins of the Oxford Polytechnic for his generous help in suggesting papers which should be included in this work and allowing me to use copyright material from his publications.

5 REFERENCES

1. Farrington Daniels, "Direct use of the Sun's Energy", Yale University Press, New Haven and London, 1964.
2. J.C.McVeigh, "Sun Power: An Introduction to the Applications of Solar Energy", Pergamon Press, Oxford, 1977, Chapter 1.
3. Ken Butti and John Perlin, "A Golden Thread", Van Nostrand Reinhold, New York, 1980, Chapter 3.
4. August Mouchot, "La Chaleur Solaire et Ses Applications Industrielles", Gauthier-Villars, Paris, 1869.
5. J.Harding, "Apparatus for Solar Distillation", Paper No.1933, Selected Papers, Institution of Civil Engineers, Vol.73, 1908.
6. W.Adams, "Cooking by Solar Heat", in Scientific American, 19th June 1878.

7. Ken Butti and John Perlin, op.cit., Chapter 16.
8. Ken Butti and John Perlin, op.cit., Chapter 10.
9. Frederick Soddy, "Matter and Energy", Williams and Norgate, London, 1912.
10. Frederick Soddy, "Cartesian Economics". Two lectures delivered at London University, November 1921. Henderson, London, 1924.
11. Commission of the European Communities, Solar Energy Research and Development, 200 rue de la Loi, B 1049 Brussels, Belgium.
12. C.Frolich, "Contemporary measures of the solar constant" in The Solar Output and its Variation, Colarado Associated University Press, Boulder, 1977.
13. "Solar Energy: A UK Assessment.", UK Section of the International Solar Energy Society, London, 1976.
14. J.C.McVeigh, "Energy Around the World", Pergamon Press, Oxford, 1984, Chapter 8, p 163.
15. H.Tsubomura, "Photochemical Conversion of Solar Energy", Proc. 1989 Congress ISES, Pergamon Press, Oxford, 1990, Volume 3, p2193.
16. G.Ciamician, "The Photochemistry of the Future", Science, 1912, 36, 385.
17. J.R.Bolton and M.D.Archer, "Chemical Conversion and Storage of Solar Energy - An Overview", Proc. 1985 Congress ISES, Pergamon Press, Oxford, 1986, Volume 3, p1843.
18. L.J.Heidt et al, "Photochemistry in the Liquid and Solid States", John Wiley, New York, 1960, p3.
19. G.Porter and M.D.Archer, "In Vitro Photosynthesis", Interdisc.Sci.Rev., 1976, 1, 119.
20. J.R.Bolton, "Solar fuels - the production of energy-rich compounds by the photochemical conversion and storage of solar energy", Science, 1978, 202, 705.
21. J.R.Bolton and D.O.Hall, "Photochemical conversion and storage of solar energy", Annual.Rev.Energy 4, 1979, p353.
22. H.Tsubomura, "Chemical Conversion and Storage of Solar Energy", Energy and the Environment into the 1990s, Pergamon Press, Oxford, 1990, Volume 3, p1430.
23. H.Tsubomura and Y.Nakato, "Photoelectrochemical Conversion of Solar Energy", Proc. 1987 Congress ISES, Pergamon Press, Oxford, 1988, Volume 3, p2908.
24. A.Fujishima and K.Honda, "Electrochemical photolysis of water at a semi-conductor electrode", Nature, 1972, 238, p37.
25 R.Memming, "Photoelectrochemical Utilization of Solar Energy", Energy and the Environment into the 1990s, Pergamon Press, Oxford, 1990, Volume 3,p1436.
26 M.G.Hutchins, "Selective thin film coatings for the conversion of solar radiation", Surface Technology,

1983, Volume 20, p301.
27 J.C.McVeigh, op.cit., 1977, Chapter 3.
28 H.Tabor, Bull.Res.Counc.Isr.,Sect.A, 1956, 5, p119.
29 C.G.Granqvist, "Solar Energy Materials: the Role of Research and Development", Energy and the Environment into the 1990s, Pergamon Press, Oxford, 1990, 3, p1465
30 M.G.Hutchins, S.M.Christie and Hu Xingfang, "Electrochromic properties of rf reactively sputtered tungsten oxide films", Applied Optics in Solar Energy 111, Czechoslovak National Academy of Sciences, 1989, p34.
31 C.M.Lampert, "Electrochromic materials and devices for energy efficient windows", Solar Energy Materials, 1984, 11, p1.
32 C.M.Lampert, "Advances in optical switching technology for smart windows", Proc.1989 Congress ISES, Pergamon Press, Oxford, 1990, Volume 3, p2183.
33 C.M.Lampert and C.G.Granqvist (Eds),"Large-area Chromogenics: Materials and Devices for Transmittance Control", SPIE Opt.Engr.Press, Bellingham, 1989.
34 C.G.Granqvist, op.cit., (1990) .
35 L.F.Jesch, "Using transparent insulation in solar energy applications", Int.J.Ambient Energy, 1988, 9, 4, p203.
36 V.Wittwer, "Transparent Insulation Materials", Energy and the Environment into the 1990s, Pergamon Press, Oxford, 1990, Volume 3, p1344.
37 K.G.T.Hollands, "Honeycomb devices in flat-plate collectors", Solar Energy, 1965, 9, 3, p159.
38 L.F.Jesch, "Conclusions", Proc.Int.Workshop on Transparent Insulation Materials for Passive Solar Energy Utilisation, Freiburg, 1986, p52.
39 V.Wittwer et al, "Translucent Insulation Materials", Proc. 1985 Congress ISES, Pergamon Press, Oxford, 1986, Volume 2, p1333.
40 W.J.Platzer and V.Wittwer, "Total energy transmission of transparent insulation material", Proc.Workshop on Optical Measurement Techniques, Ispra, Italy, 1987.
41 A.Luque and G.L.Araujo (Eds),"Physical Limitations to Photovoltaic Energy Conversion", Adam Hilger, Bristol, Philadelphia and New York, 1990.
42 M.A.Laughton (Ed),"Renewable Energy Sources", Watt Report 22, Elsevier Applied Science, London, 1990.
43 M.Telkes, "Solar Energy Storage", ASHRAE Journal, September 1974, p38.
44 A.Brandstetter and S.Kaneff,"Materials and Systems for Phase Change Thermal Storage", Energy and the Environment into the 1990s, Pergamon Press, Oxford, 1990, Volume 3, p1460
45 J.T.Beard and M.A.Ebadian (Eds),"Solar Engineering-1990", Proc. 12th ASME Conf., ASME, New York, 1990.

Information

Colour Photography

N.E. Milner

RESEARCH DIVISION, KODAK LIMITED, HARROW, MIDDX HA1 4TY, UK

1. INTRODUCTION

There can be very few people who are not familiar, from personal experience, with colour photography in one form or another, but I would like to start by quoting a few statistics which I think will surprise you. For instance did you know that an estimated 50 billion colour prints were made in 1989? If we assume that each of those prints was five inches long and they were strung together end-to-end, that string would cover 3.9 million miles. To picture such a distance is obviously difficult but for numbers of astronomic size, an analogy with astronomic distances is useful. In this particular case, you could stand on the moon and orbit around the earth 2.6 times before you would come to the end of that string of colour prints.

Of the total film market, colour is by far the biggest segment. The figures available for the USA and some European countries suggest that 88-93% of all film used is colour. In fact it is estimated that 97% of all still pictures taken by amateurs in the USA are on colour film, and 95% of those are colour negative as opposed to colour reversal.

The US is the largest single market, and in 1989 Americans took over 15 billion pictures - that is an average of over 42 million pictures every day of the year. Worldwide, the market for colour negative film has been growing at such a rate that the expected sales this year are of one thousand nine hundred and fifty nine million rolls of film, and that is twice as much as in 1980. In addition to the steady improvement in image quality, part of the reason for this growth must be the decrease in cost in real terms of films and their processing.

Let's now follow the technical milestones along the evolution of the technology which has enabled the production of these vast numbers of pictures.

2. THE EVOLUTION OF THE COLOUR NEGATIVE-POSITIVE SYSTEM

In 1826 Joseph Nicephore Niepce produced what is generally regarded as the first permanent photograph made using a camera. This was taken by exposing a thin film of asphalt on glass to light in a camera for about eight hours. The asphalt hardened where exposed and the unhardened material was removed with oil of lavender.

From about 1839 silver halide photography began to have a commercial life based on processes discovered by Daguerre and Fox Talbot. In 1841, coincidentally exactly 150 years ago, we have what I consider to be the first chemical milestone leading to modern photographic processes: Fox Talbot published details of the process of photographic development. On very short exposure to light an invisible latent image is produced in the silver halide grain. The presence of this latent image renders the whole grain significantly more susceptible to conversion to silver metal by the action of a reducing agent than an unexposed grain.

Additive Colour Photography

Even in those early days there was obviously a market for the products of colour photography, but the technology to meet this demand consisted of hand painting the available black-and-white prints. The first genuine colour photograph was produced in 1861 by James Clerk Maxwell who took a photograph of a ribbon using an additive colour photographic system, that is one involving the combination of red, green and blue images to produce the final picture.

Putting it very simply, he passed the light from his subject through a blue filter (which only allows the blue component of the light to pass) and took a black-and-white picture of this blue component. Subsequent photographic processing to produce a positive black-and-white transparency gave a picture which maps the blue light coming from the subject - black where there was no blue light and white where there was blue light. He repeated this whole procedure using green and red filters respectively to produce maps of the green and red light coming from the subject.

The original blue, green and red light distributions of the subject were then recreated by projecting each of the three black-and-white colour maps using the appropriate filter. Superimposing these projected images recreated the original colour scene.

There are several features of this process which reduce its "convenience rating" for the aspiring colour photographer, not least of which is the fact that it should never have been capable of working at all. The silver halide emulsion which Clerk Maxwell used was sensitive to ultraviolet and blue light, just barely sensitive to green light and not sensitive at all to red light. In 1961, an examination of the characteristics of the filters used in this experiment revealed that the red filter transmitted both red and ultraviolet light. This, together with the fact that many red dyes also reflect ultraviolet light, was the likely cause of Clerk Maxwell's ability to record an image of his ribbon using the "red light" reflected from it.

Other drawbacks are the need to expose and process three separate photographic plates together with the need for the parallel processing of the red, green and blue light using three projectors.

Subtractive Colour Photography

The reason for the separate processing of the red, green and blue light is that the absorber, black metallic silver, which modulates the light is opaque to all visible light. In theory at least, this problem was solved when in 1869 Louis Ducos du Hauron published an alternative method of forming a colour picture, the subtractive method. If the colour maps could be formed using absorbers specific for the light being absorbed, the light could be processed using three colour maps in series, one projector and white light.

The order of the maps is not critical but, as an example, the first colour map would modulate red light and transmit the green and blue without effect, the second would modulate the green light and transmit the blue light together with the already modulated red light while the third colour map would modulate the blue light and transmit the previously modulated red and green light. The absorbers that accomplish this are dyes of the subtractive primary colours: cyan, magenta and yellow which absorb red, green and blue light respectively.

There was, of course, no technology in 1869 which would enable this theory to be implemented in a photographic process. Not least among the problems was the very limited spectral sensitivity of silver halide, but progress was being made.

Sensitising Dyes

In 1873, Prof. H.W.Vogel accidentally discovered the effect which led to the development of spectral sensitising dyes and the ability to endow silver halide with sensitivities to the whole visible spectrum and beyond.

In simple qualitative terms, pure dyes, in very small quantities are adsorbed onto the surface of the silver halide crystal where they render the silver halide sensitive to the wavelength of light which the dyes absorb.

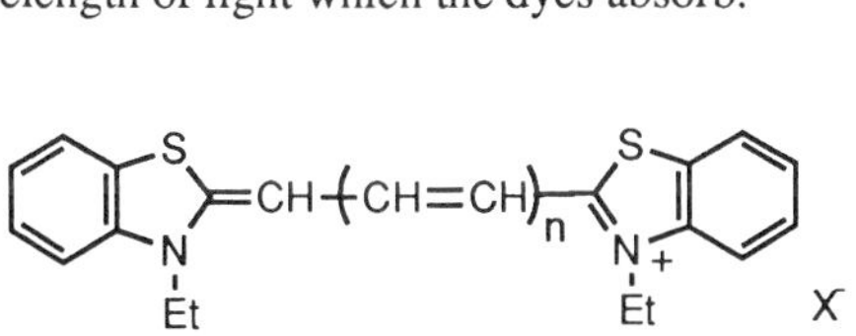

Figure 1 Structure of Benzothiazole cyanine

The cyanine dyes are a class of sensitising dyes and Fig. 1 shows a typical structure. The solution absorption spectra of the dye varies with the length of the methine chain. When n=0, the monomethine dye absorbs in the blue, when n=1 the trimethine dye absorbs in the green, when n=2 the pentamethine dye absorbs in the red and when n=3 the heptamethine dye absorbs in the infrared. When adsorbed onto silver halide the dyes enhance the sensitivity of the silver halide to regions of the spectrum which qualitatively track their solution light absorption characteristics.

The bottom line of this discovery is that we can enhance the natural sensitivity of silver halide to other regions of the spectrum. Other problems still have to be overcome but at least we have the potential to produce a system which is capable of recording information over the whole of the visible spectrum and beyond.

The Integral Tripack

The next step towards convenient colour photography is not chemical but is an arrangement which enables us to capture the three colour records simultaneously - it is called the *Integral Tripack.* It consists of three layers of silver halide coated on a suitable support, the first (furthest from the support) sensitive to blue light, the second sensitive to green light and the third sensitive to red light. The essential feature is that the layers are in optical contact and are exposed and processed as a single sheet. In order to correct for the natural sensitivity of silver halide to blue light, which is therefore a characteristic of all the layers, a yellow filter layer is included beneath the blue sensitive layer to prevent blue light reaching underlying layers.

This configuration was used by Schinzel in 1905 in a system which was a forerunner of the dye bleach process. Schinzel attempted to generate the subtractive images proposed by du Hauron by starting with layers containing silver halide and preformed image dyes - yellow in the blue-sensitive layer, magenta in the green-sensitive layer and cyan in the red-sensitive layer. After exposure and black-and-white development he proposed to bleach these dyes by the action of hydrogen peroxide in the presence of the developed silver image. Unfortunately this system did not work as the bleach reaction was not selective for the areas of silver image.

Du Hauron in 1895 proposed a similar idea to the integral tripack, which he called the *Polyfolium Chromodialytique*, but his elements were coated on individual sheets which were separated after processing. Smith had an even more similar idea in 1903: his elements were in optical contact but, again, were separated for processing.

Let us assume that we expose our tripack to a coloured subject. We now have the red, green and blue colour information recorded in separate layers. How do we convert this information into colour?

First a bit of non-photographic background. In 1881 Koechlin and Witt discovered that phenols will react with p-phenylenediamines in the presence of base and an oxidant to form a dye.

Fig. 2 shows the application of this reaction to illustrate the formation of a yellow dye from a pivaloylacetanilide, a magenta dye from a pyrazolone and a cyan dye from a phenol.

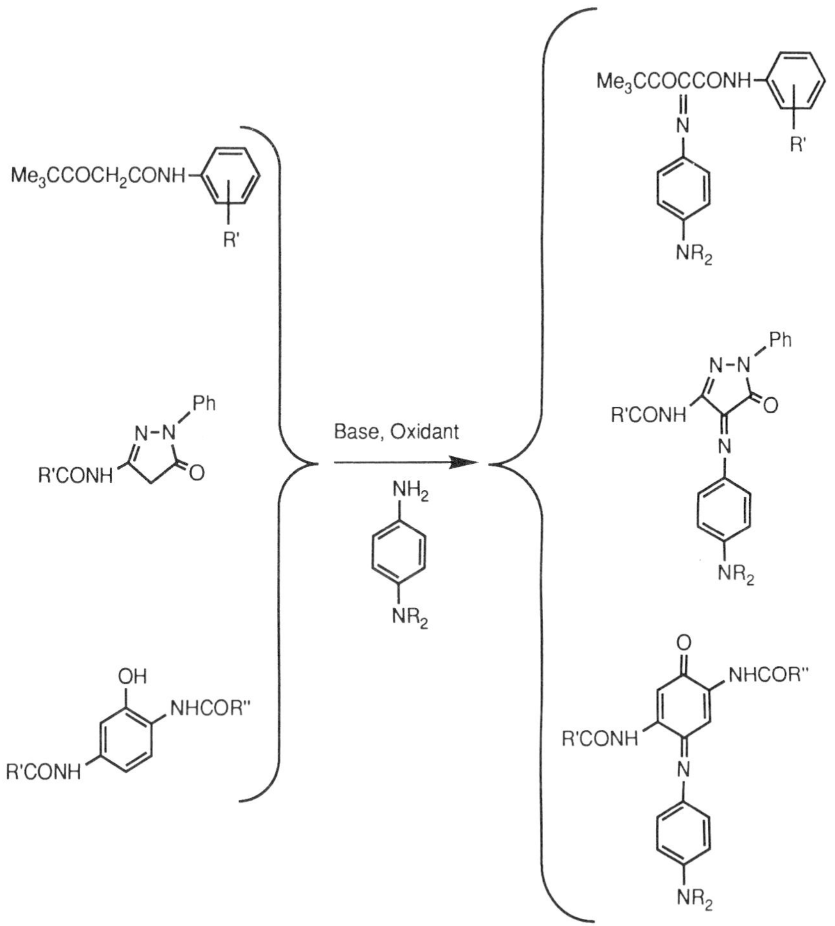

Figure 2 Equations for dye formation

Another vital piece of the jigsaw was provided by Andresen in 1888 who discovered that p-phenylenediamine could act as a developer (Fig. 3), but it was not until 1912 that Rudolph Fischer demonstrated what I regard as probably the most significant chemical milestone in the evolution of modern colour photography.

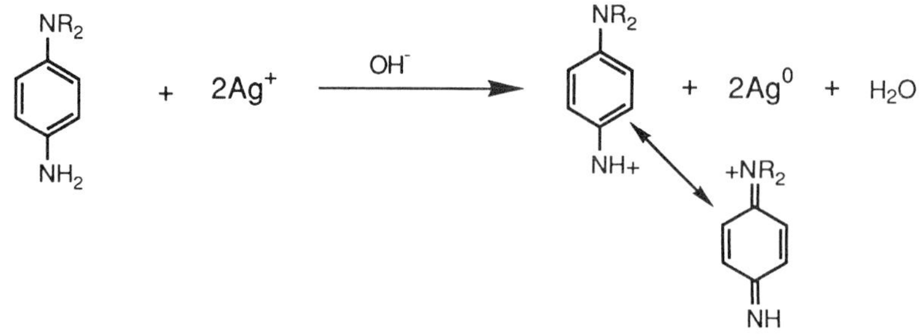

Figure 3 Development by p-phenylenediamine

Colour Formation Linked to Development

Fischer combined these independent discoveries to generate dye as a function of silver halide development (Fig. 4).

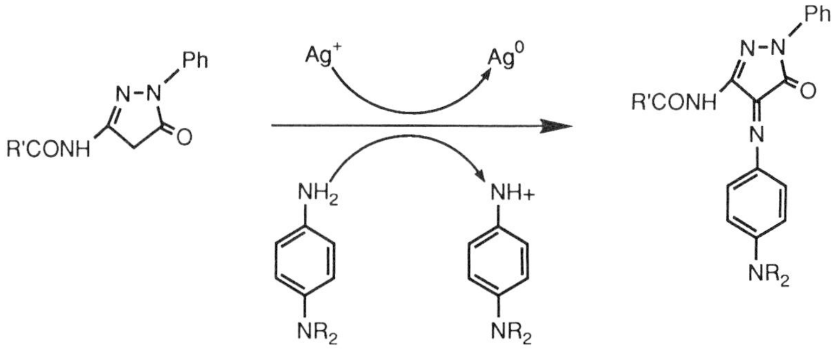

Figure 4 Colour formation linked to development

Using the pyrazolone as an example, the mechanism of the dye formation is shown in Fig. 5.

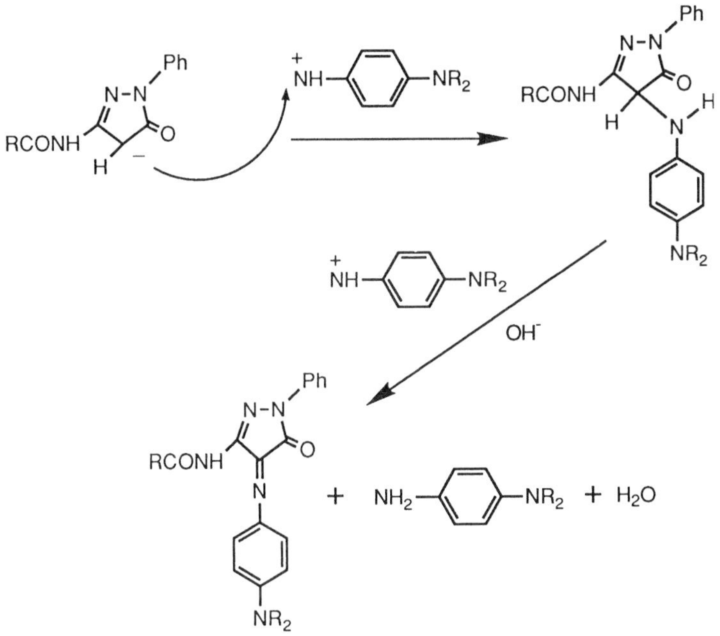

Figure 5 Mechanism of dye formation

The product that he proposed consisted of an integral tripack containing colour forming compounds, which he called colour couplers. These he incorporated in each layer such that he would form the dye complementary to the layer's sensitivity - cyan dye in the red-sensitive layer, magenta dye in the green-sensitive layer and yellow dye in the blue-sensitive layer - as a function of silver halide development.

This process should, in principle, have been capable of producing the subtractive colour maps proposed by du Hauron. The problem was that Fischer could not ensure that the colour couplers remained in the appropriate layers. The result of the interlayer movement was gross colour contamination of the final colour image.

It took until the mid 1930s for photographic manufacturers to find ways of overcoming this problem. The products were the colour reversal transparency films *Kodachrome* and *Agfa Neu*. Each of these products solved the problem in a different way.

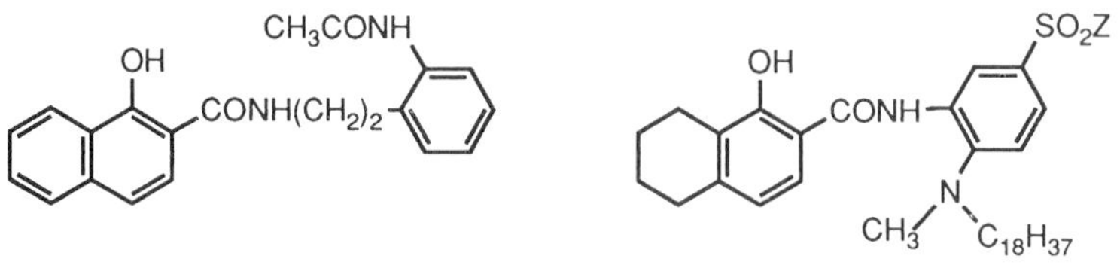

a) Developer-Soluble Coupler

b) Z = OH Micellar Water-Soluble Coupler

c) Z = $N(CH_3)_2$ Oil-Soluble Coupler

Figure 6 Types of coupler

Fig. 6a shows a coupler of the type used in *Kodachrome* film and which is designed to be soluble in the developer solution. The actual photographic product contains no coupler but during processing the layers of the tripack are developed sequentially in a developer containing the appropriate coupler. Fig. 6(b) shows a coupler of the Agfa type which is incorporated in the way that Fischer intended. It is designed to be non-diffusing by the inclusion of the long chain hydrocarbon group and is soluble in the same way as a surfactant. The non-diffusing properties were explained using the analogy of "walking through a dense forest carrying a tree".

Fig. 6(c) shows a structure which solves the problem in yet another way. The coupler is designed to be oil-soluble and is incorporated into the appropriate light-sensitive layer in the form of tiny droplets of coupler-in-oil solution. This method, which was first demonstrated in *Kodacolor* film when it was released in 1942, is currently used in most colour negative films.

We can summarise the steps in the implementation of a generic colour negative/positive process as follows:

(i) Exposure of the film to a subject. This generates a latent image in the silver halide of the appropriate light sensitive layers.
(ii) Colour development of the exposed silver halide and generation of dye by reaction of the oxidised colour developer with the coupler incorporated in the layer.
(iii) Oxidation (bleaching) of the developed silver metal to a silver salt.
(iv) Removal of all silver salts (fixing) from the material by use of a suitable complexing agent (fixer).

The result at this stage is a transparent dye picture (colour negative) which represents the original red, green and blue colours of the subject in terms of their subtractive colours cyan, magenta and yellow respectively. The dye image on this colour negative is then used to control the red, green and blue exposure of a similarly sensitised piece of photographic material, usually coated on a white paper base. As this operation is usually achieved by projection, an enlargement of the original negative is possible by means of suitable optics. Repetition of the processing cycle described above produces the colour positive picture.

Coupler Technology

The picture quality that everyone, quite rightly, expects from today's colour photographic materials has been achieved by technical advances in many areas of photographic science. Certainly not least among these areas is coupler technology where a modification to the structure of the coupler results in the ability to release photographically useful compounds in an imagewise fashion.

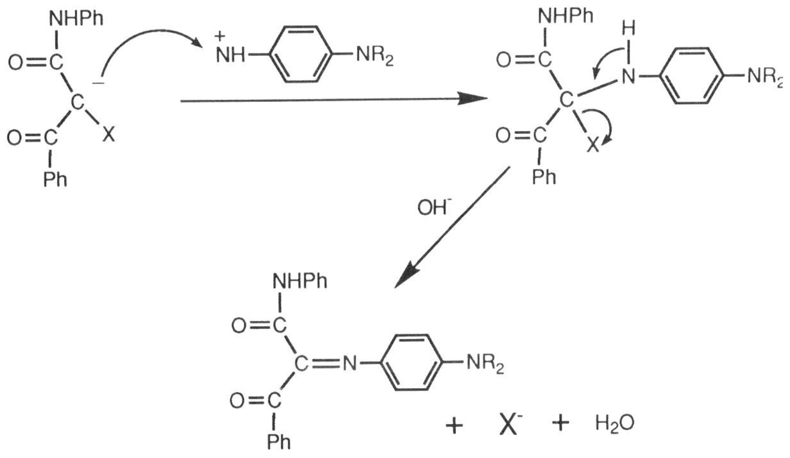

NHPh O=C C X O=C Ph + NH NR2 → NHPh O=C C H N NR2 X O=C Ph

OH⁻

NHPh O=C C=N NR2 O=C Ph + X⁻ + H_2O

Figure 7 Couplers releasing useful groups

This structural modification is shown in Fig. 7 where the coupler has a leaving group fragment attached to the position where coupling will take place. This molecule is perfectly stable until attacked by the oxidised developer but after the initial attack the intermediate eliminates the leaving group to form the dye. The effects that can be achieved by the release and migration of these fragments obviously depends on their nature.

This technology was used in the first major step in the improvement of the colour quality of the final print. This was achieved in 1948 by the use of an Integral Colour Mask in the negative material to compensate for the fact that the dyes being formed did not absorb only in the area of the spectrum that was required. A coloured colour-forming coupler, which carried a dye function as a leaving group, was used. The loss of this group by the coupling reaction destroyed the original colour while generating the image dye colour. By arranging that the original colour of the coupler was the same as the unwanted absorption of the dye being formed the combined effect is a uniform "unwanted absorption" over the whole negative. Compensation for the effect of this uniform absorption can easily be applied during the printing of the negative.

Perhaps one of the most versatile leaving groups is the development inhibitor. Derivatives of mercaptotetrazoles and benzotriazoles are common examples of development inhibitors which are innocuous when suitably attached to a coupler but when released they inhibit further development of the silver halide. This is a very significant achievement as the controlled diffusion of these development inhibitors, after release, both within and between the imaging layers before they inhibit the development of a silver halide grain, enables the photographic scientist to control and improve such diverse properties as colour saturation, sharpness and graininess in the final print.

3. SOCIAL AND ENVIRONMENTAL IMPACT OF COLOUR PHOTOGRAPHY

Social Impact

For many people colour photography is associated with the snaps which they take of their families and friends at home or on holidays. These pictures record special moments but have you noticed how other people's holiday snaps are never as enthralling as your own?

The appeal is very personal because the images bring an occasion back to life. The photographer and the subjects relive that occasion and all the feelings that went with it whenever they look at the picture.

There is of course much more to colour photography. Its uses are prodigious and enter our lives in so many ways. Photography is about communication and if a picture is worth a thousand words, then how much more might it be worth in colour? We use photography to record, to educate, to inform, to explore and to entertain. When we think of entertainment in this context, we think of the movies.

Whether it be romance, adventure, terror, political intrigue or pure escapism, movies make waves in our culture. They bring us a close encounter of a very special kind and become part of the world culture, crossing language boundaries and every other sort too.

Commonly, for the period that you are watching a film, the movie maker wants you to believe in the reality of the scene that he is setting before you and features, such as colour, that enhance this sense of reality are obviously important. For a milestone in the movies we need look no further than *Gone With The Wind.* This was such a blockbuster that it might have been just as successful in black-and-white although it's my opinion that Scarlet in black-and-white would not have had the same appeal!

From all over the world the media bring colour pictures into our homes to make us aware of human conflict and suffering, and of environmental issues. Let's now look at some aspects of how photography interacts with the environment.

Direct Environmental Impact

Photography and silver go hand in hand. For 150 years, no-one has found a substitute to equal the performance of this precious metal in photographic systems.

Table 1 Silver Consumption Estimates for USA

	Photographic consumption	% of total manufacturing consumption
1990	68.0 m oz	55.3%
1980	49.8 m oz	39.8%

The figures shown in Table 1 for the consumption of silver by the photographic industry in the USA illustrate the importance of silver. Note the 15% rise in the proportion of silver consumed by photography in the USA in the last ten years.

The amount of silver that is recoverable from photographic materials varies a lot for different products and types of waste. As we're talking about colour photography today, let me remind you that there is no silver left in any of your processed prints and transparencies because the colour process produces a final dye image from which all silver has been extracted. Silver may be recovered in high yield from colour processing solutions. In practice the amount that is reclaimed varies according to the process and the application. Trade estimates are that it is at least 60% and in some cases over 90%. Potentially even greater recovery is possible.

Indirect Environmental Impact

Let me now turn to examples of how colour photography enhances our awareness of what's happening to and in our environment. One way in which we can get a much wider perspective of our environment is to use aerial photography in combination with colour films.

For example conventional aerial colour photography is an effective means of monitoring pollution in our rivers and lakes but a striking way to contrast different features in the environment is to use false colour infrared photography. This picks out features which have a high reflectance in the near-infrared portion of the spectrum; they are imaged red on a colour infrared photograph.

Different types of vegetation record as different shades of red but more importantly, plants under stress lose some of their infrared reflectance and record in other colours. This effect is used extensively in forestry and agriculture as a means of disease detection in vegetation. It enables you to spot a diseased tree a mile off, literally.

While on the subject of trees, an environmental issue which National Geographic Magazine has illustrated by the effective use of colour pictures is the destruction of tropical rainforests. Pictures showing the ravages inflicted by large-scale mining operations, whether for minerals or trees, provide very effective contrast to idyllic images of unspoiled forest, exotic vegetation and endangered species of animal. The message that "something worth preserving is being lost" is graphically driven home.

In fact for the greater part of this century the photographers of the National Geographic Magazine have gone to almost any depths to bring the wonders of the world to our attention. In 1926 two of them obtained the first colour underwater pictures using photographic (Autochrome) plates hypersensitized with mercury and a "reflector raft" to compensate for the low light conditions under water. Their expedition included a supply of flashpowder so that when the man on the surface got a signal, via a piece of string, from his colleague 15 feet below to alert him that a fish had appeared in front of the lens, he detonated a pound of flashpowder beneath the raft. Underwater, this provided the required level of illumination to obtain the historic pictures; on the surface, the effect was most definitely not for those of a nervous disposition.

4. CONCLUSION

From its first faltering footsteps over 150 years ago, we have followed the evolution of one branch of colour photographic technology taking note of some of the major milestones which mark the way. These milestones and those of the colour photographic systems which time has not permitted me to mention, have been placed by scientists and inventors showing a degree of innovation, ingenuity and perception which it is difficult to match in any other field. These qualities have generated a technology which can provide the superb quality pictures which everyone today, quite rightly, takes for granted.

Acknowledgements

The author gratefully acknowledges the help and assistance of Miss J. Menton and other colleagues during the preparation of this lecture. The cooperation of James P. Blair of National Geographic Magazine and Volkmar K. Wentzel is appreciated.

Materials for Electronics

W.O. Baker

AT&T BELL LABORATORIES, MURRAY HILL, NJ 07974, USA

The unity of nature seems to be well embodied in the electromagnetic behavior of matter. (Perhaps though, because of our ignorance of life forces, we are being naive as to whether there is some other principle, apart from thermodynamics mass/gravity, of comparable meaning.) Nevertheless, the relations of electricity to materials are relatively new ideas in human history, even though the Greeks had a word for electricity and found electrostatic repulsion and attraction in some kinds of matter, especially insulators, and magnetism in other kinds of matter, especially certain minerals. The development of modern "electronic" materials has thus followed the basic knowledge of materials -- parallel to the evolution of atomic physics and of chemistry.

The Greeks also thought about atoms and chemistry, but the electrical nature of atoms is a new notion. It was just 94 years ago when Sir J. J. Thomson postulated and demonstrated the existence of the electron. Accordingly, in any specific or modern sense, the intimacy of electricity and chemistry with its essence of valency, of bonding, and of reactivity through electrons is new. Such a link has existed in specific forms over a time span of only twice the number of years that many of us here have been working at the understanding of it.

Indeed, for the last half century, the progress of quantum mechanics and Davisson's and Germer's proof of the wave nature of the electron have turned the essence of chemistry toward describing the locale and energies of electrons -- such as their spin quanta, their localization or delocalization in valency, etc.

Likewise, in electromagnetic terms, the connection of magnetism with electricity came first from the work of H. C. Oersted in 1820, despite the observations millennia before on the magnetism of iron minerals and the invention of the compass, presumably by the Chinese. So the theme of our report is the confluence in developing condensed matter of the concepts and sciences of chemistry and electromagnetism. Specifically, it reflects some direct experiences with how principles of chemistry have been part of the strategy of research and development which achieved in industry the modern electronic communications and computer eras. This convergence came beautifully from the self-taught intellect of Michael Faraday (1791-1867). Faraday, given a loaf of bread as nourishment that had to last him through a week in the time before his father's death in 1809, passed on to humanity the loaves of learning, the bread of our modern electronics technology and culture, which sustain us in this century.

For Faraday first grasped a unity of the physical science of matter and energy, a unity in which chemistry and electricity are prime ingredients. So, about 178 years ago in March when Davy brought Faraday into the Royal Institution as his laboratory assistant, there began the great train of bold discovery. From it has arisen nearly every feature of modern chemistry, electricity, and indeed, in the post-Newtonian phase, much of optics, especially what we call photonics. By the summer of 1832, a century and a half ago, Faraday supported the "wild surmise" that electricities coming from chemical voltaic cells, electrostatic generators, thermocouples, dynamos, and electric fishes were identical. Has there been a more momentous idea -- at least before Einstein and the interconvertibility of matter and energy, or before Heisenberg and the quantized atom?

Now, when the Royal Society of Chemistry draws us together, we find that materials, especially in the solid state, are strong factors in the 20th Century resources of electronics.

Emergence of Vacuum Electronics

This role of materials shifted strikingly after Lee deForest invented the audion, a three-electrode tube, in 1906 and proposed it for long-distance telephony in 1912. After that H. D. Arnold of the Western Electric, and Irving Langmuir of the General Electric Company, discovered the effectiveness of high vacuum for electron circuit amplification. Manufacture of these electron tubes was begun by Western Electric 77 years ago, with transcontinental telecommunications enabled thereby early in 1915, and transoceanic radio telephony coming on in 1927. Thus emerged ever-rising needs for materials development to sustain the high temperatures and vacuum stability of the electrodes and structural components in these earliest tubes. These supported the increasing circuit stability of the total networks and systems, cables, antenna, and other crucial elements through which these electronic signals had to move. Until then, heavily responsive and cumbersome systems had characterized the age of electricity and magnetism--batteries, with their electrolytes, massive electro- mechanical generators and motors, relays and carbon microphone transducers--all depending on conventional properties of conductors. Such conductivity ranged from about .8 reciprocal ohm centimeters for the most concentrated battery acids downward to 10^{-3} to 10^{-4} reciprocal ohms centimeters for weak electrolytes and upward to about 2×10^6 mhos for familiar metals at ordinary temperatures. The conductivity of graphite at ordinary temperatures at 10^{-3} mhos is an intermediate (and metallic) figure. For contrast, early electrical insulators, such as glass and plastics have resistivities 10^{14} to 10^{16} ohm centimeters which at relative humidities of 90% often fall to 10^7 to 10^9. So along with environmental sensitivity a spread of 10^{20} in specific resistivity between the metal conductor and the nominally ideal nonconductor, accents the range of materials properties in electrotechnology. After all, it is the inner electronics of matter, from the completely localized atomic orbitals of a perfect insulator to the band structure of the metals with an easy electron mobility, that define this fabulous quality covering over 18 to 20 orders of magnitude. These materials properties concurrently sustain the circuitry and technical functions of the systems of signal handling in telecommunications and computers, along with the sensors and transducers, all of which are prime purposes of electronics.

Combined Materials Properties for Electronic Subsystems.

Likewise, of course, these components comprise the states of physical tangibility and stability whose dimensions enable huge phased array radars and ultramicroscopic integrated circuit chips. So their development interacted strongly also with obtaining exacting mechanical properties of the conductors and dielectrics, especially in the last half century beyond the vacuum tubes and into the networks of solid state and semiconductor science and technology. Obviously, in between the extremes of ultimate insulation and high conductivity lie also the dramatic qualities of the semiconductors, whose resistance in ohm centimeters was long ago measured as 10^6 or 10^8 in comparison to the 10^{16} characteristic of the purely insulating structures such as paraffin, or to the metal conductors of 10^{-6} resistance. Casually inserted in some of the older tables of resistivity are these intermediate figures, such as 2×10^{-1} for the element tellurium.

Significantly, in thinking of the simplicities of electronic conductors and nonconductors, a combination pursued for a half century of personal effort, one of the historic successes of polymer chemistry and electronic materials had its origins in the United Kingdom. For polyethylene (first applied in 1935)[1] turns out to be the ultimate and practical dielectric for high-frequency electronics, like microwave systems for radar and telecommunications.[2] Correspondingly, these applications underlie fundamentals of polymer chain structure through the influence of polar substituents, stability to oxidation and other chemical modification and physical behavior, such as in insulating cores of coaxial and transoceanic cables exposed to moisture on other polar penetration. So the dielectric constant of 2.33 of relatively pure polyethylene hydrocarbon, with its independence of frequency, and the attendant specific resistivity of 10^{18} ohm centimeters, mark a take-off point, as well as symbolic diversity, of how 20th Century electronics have involved so much of modern materials and solid state chemistry.[3 4] The superbly low dielectric loss of polyethylene, tan δ, and relative immunity to water, even under pressure, reflect in Fig. 1 the new era of insulating materials demanded by electronics. These are accompanied by mechanical strength, as shown in Fig. 2, which depicts the discovery of sensitivity of microcrystalline polymers to biaxial stressing intrinsic to coverings of wires and cables.[5]

The interface of electronics with materials has actually provided a vast array of other examples, analogous to this role of dielectric polarization, in chemical structure and materials qualities and utilization. For instance, conventional concepts of molecular and solid state structure and chemical binding have been extended by the prominent, indeed central, concepts of charge transport by positive holes, p, as well as electrons, n, and of dislocations and defects as well as solute donors and receptors, as special electronic parameters in materials.

Pivotal Role of Crystal Formation

Such mechanical, chemical and electrical properties reflect the astonishing process of crystallization, in which myriads of particles nearly simultaneously arrange in surfaces that solidify with precise positional regularity. It turns out that this technically trite remark contains the major motifs in the development of electronic materials. And, coordinately, these findings have disclosed crystal qualities that affect virtually all technology-- from strength of matter, to reaction catalysis, to cellular habits of proteins and viruses.[6]

Accordingly, many computer models of crystal formation affirm that atoms get misplaced -- indeed, probably couldn't aggregate extensively, without screw dislocations, resulting vacancies, etc.[7] This appears in computer-generated diagrams in which known properties of crystallization from the melt are used in suggesting causes of imperfections, as in Fig. 3. Nature's mixing of elements has also injected impurities at formerly unmanageable levels.[8] Exposure to air and other agents likewise insures surface changes.

This over-all perspective in the development of electronics has influenced various features of the physics and chemistry of matter. Pfann's application of zone refining by moving a liquid zone through a solid ingot (first with Ge as seen in Fig. 4, and thus applying the liquidus/solidus phase relations of impurities, has produced Si with less than 10^{17} C and less than 10^{15} other elemental atoms/cm^3 (many orders more pure than before achieved in studies of this planet's prevalent element Si). Single crystals[9] treated as solutes in Fig. 5, governed by the distribution coefficient k, yields pure, nearly perfect products. Instead of the typical dislocation density 10^8 or more per cm^2, for most crystals (including metals), Si has been obtained essentially dislocation-free, although technically useful forms have $3.5 \times 10^6/cm^2$; and even $10^3/cm^2$ dislocations were shown to affect mechanical yield and strength.

Indeed, the original experimental demonstration of crystal dislocations predicted by solid state theory[10] was in zone refined Ge delineated by the etch pit pattern of Fig. 6.[11] Nowadays, the chemistry (catalysis, etc.) as well as mechanics (strength of rockets, aircraft, bridges, etc.) of solid matter involves this earlier knowledge of dislocations and imperfections, as well as of newly determined purity.

Although this sector of electronics-induced materials established the non-ideality and artificial modification of solid matter, there does seem to be evidence of virtual perfection in nature's metal "whiskers", discovered around electronic circuit elements.[12 13] Electric fields promote ion and other migration. Electroplated Zn, Sn and Ca yield filaments about 0.6 cm long and 3 μm in diameter having tenacity and yield stress indicating nearly ideal physical properties--far beyond conventional values. This has led to many fiber matrix composites, for aerospace-applications. Such "natural" growth of Fe whiskers is shown in Fig. 7. But controlled synthesis of Si whiskers was also an early exercise in Vapor-Liquid-Solid synthesis for electronics.[14]

Still other phase rule adaptations besides zone refining have come from electronics technology. For instance, the versatility of disequilibrium aggregates seems ready for new tracks, as appears in Plewes' discovery of spinodal decomposition in bronzes, whose

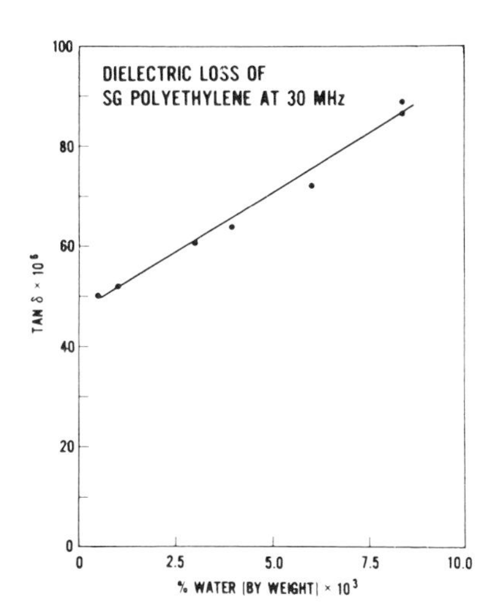

FIGURE 1. EFFECT OF WATER ABSORPTION SUCH AS IN SUBMARINE CABLES AT GREAT DEPTHS ON THE DIELECTRIC LOSS TANGENT δ OF HIGH PERFORMANCE POLYETHYLENE. ILLUSTRATING INSULATION QUALITY IN PPM UP TO 1/10,000% OF THE HIGHLY POLAR WATER MOLECULES.

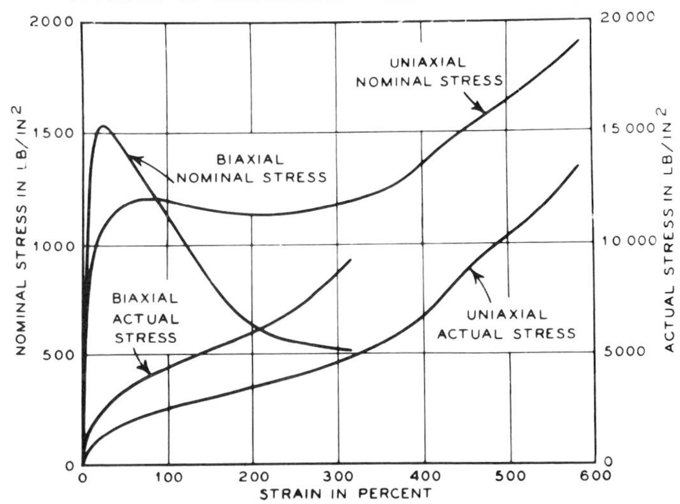

FIGURE 2. MECHANICAL PROPERTIES OF POLYETHYLENE UNDER SPECIAL MULTI-AXIAL STRESS ENCOUNTERED IN CRITICAL ELECTRONIC INSULATION IN CABLES, WIRES AND OTHER CIRCUITS. ANALYSIS OF THE POLYMER SOLID STRUCTURE LED TO SYNTHESIS OF A STABILIZED POLYMER PROVIDING MANY DECADES OF SERVICE.

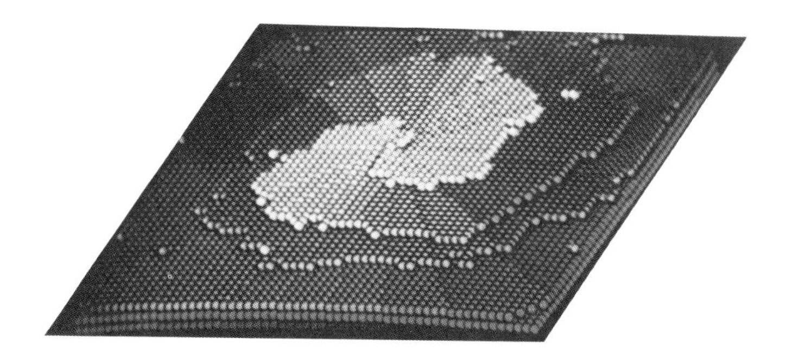

FIGURE 3. MODEL OF CRYSTALLIZATION INVOLVING SCREW DISLOCATIONS FOR INITIATION OF ORDERED SOLIDS, ACCORDING TO COMPUTER ANALYSIS OF PROBABLE CRYSTALLIZATION KINETICS.

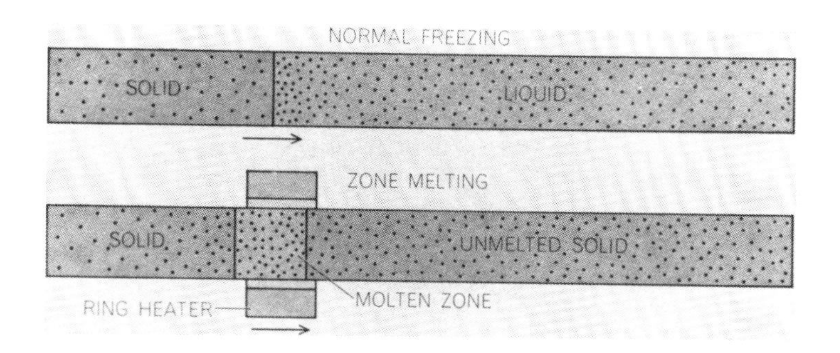

FIGURE 4. SCHEMATIC OF THE ZONE MELTING REFINEMENT OF AN INGOT SUCH AS GERMANIUM. REPEATED PASSES OF THE MOLTEN ZONE SWEEP OUT SELECTED IMPURITIES AS THE MELTED PORTION MOVES ALONG LEADING TO UNSURPASSED CHEMICAL PURITY AS WELL AS ULTIMATELY PHYSICAL PERFECTION.

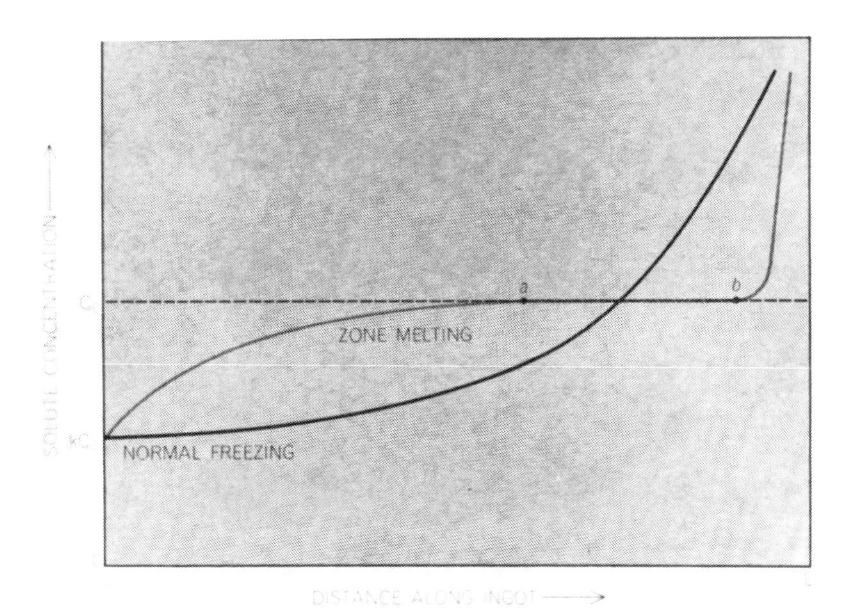

FIGURE 5. PHASE RELATIONS SHOWING SHIFTS IN SOLUTE CONCENTRATION CONTROLLED BY ZONE MELTING IN COMPARISON TO NORMAL FREEZING. IN CONCENTRATION GOVERNED BY THE DISTRIBUTION COEFFICIENT k, PHASES ARE EXCEEDINGLY UNIFORM AS WELL AS CONTROLLABLE. SOLID SOLUTIONS CAN BE PRODUCED YIELDING THE ESSENTIAL DOPING OF ELECTRONIC MATERIALS.

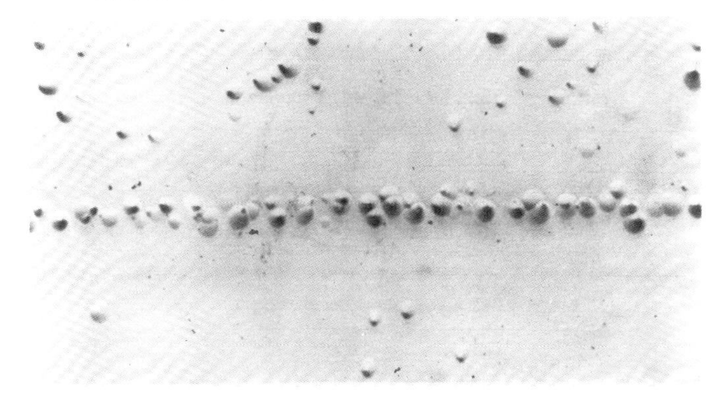

FIGURE 6. ETCH PITS IN A SINGLE CRYSTAL OF GERMANIUM PRODUCED BY ZONE REFINING AND REPRESENTING EXPERIMENTAL EVIDENCE FOR POSTULATED ATOM IMPERFECTIONS IN CRYSTALS.

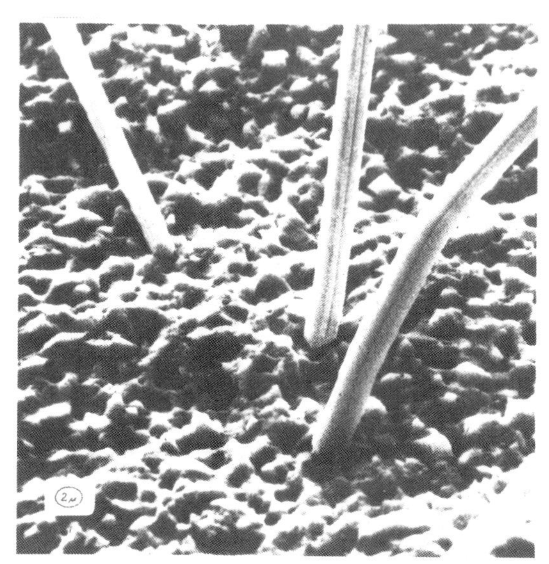

FIGURE 7. SPONTANEOUSLY FORMED SINGLE CRYSTAL WHISKERS OF IRON POSSESSING HIGH MECHANICAL STRENGTH AND FREEDOM FROM DISLOCATIONS.

physical properties were nominally optimized 1,000 years ago. Yet requirements for electromechanical supplementation through modern circuit relays led to a tripling of yield strength of the conventional Cu-9Ni-6Sn15 alloy, up to 150,000 psi yield, as seen in Fig. 8. Traces of Nb given even better properties.

Electronic Categorization of Matter

As noted, the primary driver in electronic materials has been shaped by theory and concepts of both physics and chemistry, to be realized in some practical substance, usually crystalline. This required the refinements so far described, as we now approach the electronic behavior itself. The marvelous electromagnetic property span from insulators to conductors challenged the theory for interconvertibility of insulator $\rightarrow$ conductor states. What sort of chemistry/physics would exhibit a continuum, so that we could simply pick some electronic circuit properties out of a sliding scale? This question was illuminated by a series of phase transition/critical state/quantum statistical theories advancing understanding of all solids. A particular resolution of the metal-insulator transition issue came from Prof. Sir Neville F. Mott,[15] with experimental confirmation with oxides of V, as shown in the phase diagram for pressure/temperature. A change in inter atomic spacing by external pressure causes discontinuities in electron correlations. Such striking effects appear in mixed Ti-V-Cr sesquioxides, as shown schematically in Fig. 9.[16] The percentages at the top refer to the Cr content in the V oxides. Buildup of the Ti portion similarly leads toward the "metal" phase.

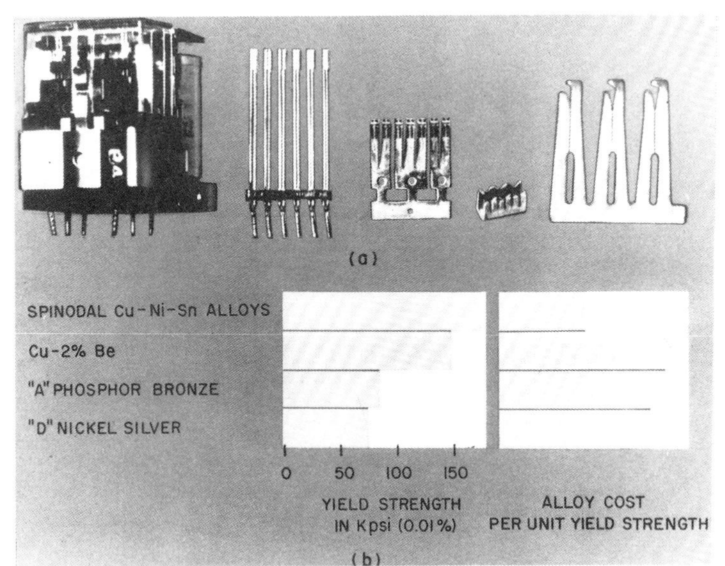

FIGURE 8. EXAMPLES OF BRONZE RELAY COMPONENTS FOR ELECTROMECHANISMS IN ELECTRONIC SYSTEMS WHERE SPINODAL PHASE PROCESSING HAS TRIPLED THE CLASSICAL YIELD STRENGTH OF BRONZE.

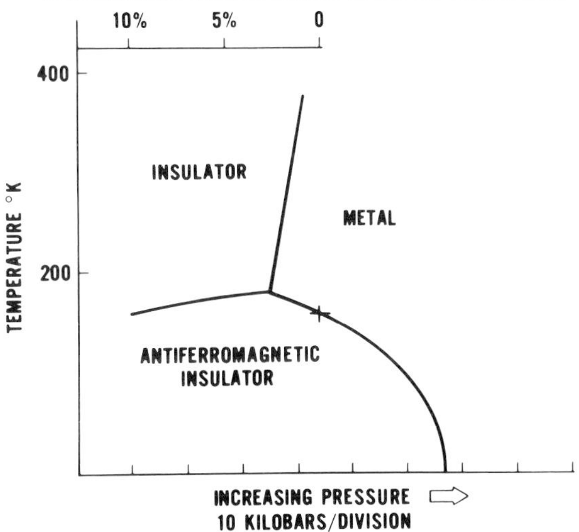

FIGURE 9. EXAMPLE OF THE METAL-INSULATOR TRANSITION SUCH AS IN VANADIUM OXIDES, EXPRESSING THE UNIFIED CONCEPTS OF CONDUCTORS, INSULATORS AND MAGNETIC SOLIDS DEPENDING ON INTER-ATOM SEPARATIONS AS EFFECTED BY EXTERNAL PRESSURE.

Now such precepts of the bulk comportment of matter supported the ultimate goal of using the atomic charges and structures for circuitry. The theoretical principles again guided our evolution of materials. Distribution of the charges among chemical entities had to be formulated.[17] But already F. Bloch had reported an inspired insight from quantum mechanics about how electrons could behave in crystal lattices.[18] The concept was next advanced by A. H. Wilson's theories of energy states, comprising valence bands (filled) and conduction bands (empty), depending on their crystal environ.[19 20] Si and Ge are of course the classic examples of valence band semiconductors, each atom of the lattice having 4 covalent bonds. So thus was built a base for intensive later studies of the correlation of physics and materials, and discovery of the transistor.[21 22]

Semiconductor Qualities

Indeed, the remarkable concordance of these physical, chemical and engineering ideas, including careful evolution of common languages, have invigorated the entire mainstream of electronics materials.

Thus the intermediate semiconductors, classically germanium and silicon, but including many others, were found to be media for new chemical units providing for not only electrons, holes, and lattice vacancies, but also dilute solutions in the crystal. The latter were vigorously pursued, especially with bonding variations such as electron donors (arsenic in silicon, using its n-type conductor valency of 5), and acceptors, (boron with valency of 3, offering p-type conductivity). The pathways chosen for these compositions is illustrated in Fig. 10, where B is "dissolved" in the Si lattice, and Li is inserted in interstitially, as a donor.

So here appears the wide frontier of materials development,[23] which has enabled

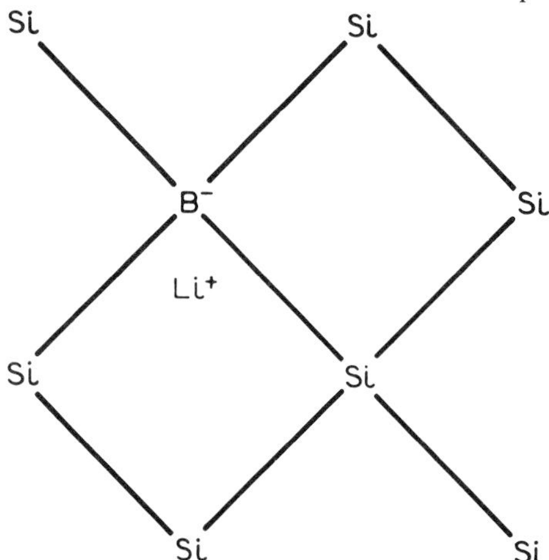

FIGURE 10. SCHEMATIC OF CHEMICAL INFLUENCE OF CHARGE DISTRIBUTION AND CARRIER SUPPLY IN THE HOST LATTICE OF SILICON.

invention by providing the environ for electric charges, which control waves and signals, thus creating a Communication and Information Age. So, as noted before, the principles of quantum mechanics, described for instance, the electrons in a perfect crystal of diamond, as inhabiting the 2s and the 2p states. These have in the 2s band 4 electrons per atom, yielding the valence band separated from the 6 1s states of tighter bonding. Such concepts were as central in our research planning as the chemical purification itself. Likewise, in the other extreme of conductivity, the metals, care was taken to understand the electrotechnical significance, such as that in the element sodium. Its bands of the s and p levels are not so decisively separated as in diamond. The solid state implications were heeded of the 11 electrons from each sodium atom being distributed with 10 going into 1s, 2s and 2p bands, filling them completely, and the remaining occupant of the 3s band coming out half filled. In still other cases, the valence band might be just filled, but in all such metals, the bands do overlap, so the solid accordingly supports what is often modeled as a condition of free electrons with high conductivity. The critical attention was focused in between, in the energetics of these solids, to transitions through the energy gaps, activated by heat, thermal agitation, or light causing photoconduction.

These general concepts, originated by A. H.Wilson,[20] were vigorously extended both theoretically and operationally, and steadily impacted much materials development. Thus, in considering the limited examples in the following, it should nevertheless be recognized that most of the principles of modern inorganic chemistry, and increasingly organic structures, are represented in the electronic materials solids. So we believe that the practical substances evolved require purity and consistency of structure, apparently unexcelled in the entire history of chemistry. Further, the idea of chemical synthesis by crystallization, in which the formation of intrinsic lattice bonds and those of added dopants or elements such as arsenic, lithium, boron, in germanium and silicon and others, has provided an extension of the ideas of chemical identities. Namely, 10^{16} or less atoms of an adduct per cm^3 containing 10^{20} atoms of the host is nevertheless a defined and manageable chemical additive. Indeed, it is such quantities which will determine the conductivity $\sigma = \frac{1}{\rho}$ (the resistivity) $= q(\mu_n n + \mu_p p)$, where $q = 1.60 \times 10^{-19}$ coulombs, μ_n = electron mobility expressed as $centimeter^2$ per volt second n is the number of conduction electrons, μ_p = mobility of holes, and p = the total number of holes. Hence, chemical control by the composition in environ of the crystal or the film establishes the electronics of the system.

Electric Charges in Semiconductors

The creation of this sort of solid solution (as in Fig. 10) can thus be a combination of a classic pure solvent and added elements as solutes. But, obviously, such solutes can be imperfections in the crystal, as well as vacancies and other defects. Depending on the energy gap separating the valence and conduction bands as noted, large numbers of compounds, as well as elements, can qualify as semiconductor media, depending on temperature and composition. These range from the most fully covalent structures like irradiated diamond through to a huge population of ionic crystals. Some compounds have high electron mobilities, μ, and very low-energy gaps, such as InSb, which is outstanding with μ of about 10^5 centimeters^2 per volt second. And even with an energy gap of about 1 electron volt, μ in germanium and silicon runs between 10^3 and 10^4. However, ionic crystals may have values of μ as low as 10 or 10^2. The hole mobility is about 1/2 or down to 10^{-2} of the electron mobility, and so we see a broad range of these media properties which have guided the development of electronic materials, especially from Periodic Table Groups II-VI, III-V and IV. In some ionic media, the number of charge carriers can be considered to have high values of 10^{22} per centimeter^3, and others can be brought up to 10^{19} ions per centimeter^3 by purposeful imperfections. Nevertheless, electron density and mobility may be less than the ion density and mobility, and there is a wide range of resultant conductivity. A consequence is that various compounds are selected for particular responsiveness, such as copper oxide, selenium and silicon for use as current rectifiers, and lead sulfide, lead telluride, selenium, cadmium sulfide, and silver oxide and cesium for photocells, (along with the especially efficient diffused silicon Bell cell).

Temperature sensitive resistor functions and circuits employ germanium, boron and uranium oxide, U_3O_8, whereas nonohmic resistors or varistors have come from silicon carbide. Phosphoric materials for display use zinc sulfide, cadmium sulfide, zinc oxide, and so on, whereas the cathodes of vacuum tubes discussed earlier gain electron emission and conduction by proper blends of barium oxide and strontium oxide. Particularly precise and ingenious use of mixed oxide semiconductors (Ni(20) with Mn(80) wt. percents) yielded negative temperature coefficient thermistors. These function in many systems ranging from infra-red detectors (including weapons in WW II) to frequency control in telephone signaling. Other chemical preparations of $BaTi0_3$ provide positive temperature coefficient thermistors by trapping 0 along crystallite boundaries. These ions yield an insulating layer so that when the temperature is raised above the Curie transition of $BaTi0_3$, resistivity increases by 10^4 to 10^7 times! The Curie transition can itself be adjusted over a wide range by chemical composition control.[24] Counters for high-energy particles that cause scintillation employ organic semiconductors, with their particular large-energy relationships such as in anthracene and stilbene.

Thus, it is seen that such functional properties, coupled with the general theory we have cited, have stimulated and guided the chemical development of a galaxy of electronic materials. Other features include a roster of imperfections and stoichiometry in oxides, especially of unfilled d-shell elements, and have many correlates in modern chemistry. Especially, the sensitive electromagnetic characterizations have revealed the non ideality of many, if not most, common substances, and especially in their prevalent oxide forms. Further, both physical as well as chemical properties of these substances are notably affected by the same series of imperfections that so drastically influence their semiconductor behavior.

Films and Surfaces for Perfecting Electronic Media

Examples of these interactions take a host of forms, but some of the most effective again appear in electronics technology, where surface physics and chemistry soon induced the building of new and satisfactory surfaces, rather than depending on the usual product of natural solidification or conventional etching, heat treating, etc. This is not to say, however, that novel forms of heat treating are unused, for the laser annealing[25] schemes are so effective and selective that heat waves are generated and disposed by outer electron excitation without causing significant nuclear shifts. Thus special electronic configurations are formed, useful for circuit behavior. However, this more general tactic of building new surface has generally been obtained by epitaxy[26] with almost endless combinations. An early design is liquid phase epitaxy[26] (used for GaAs) which was thence extended by the ultimate precision and versatility of molecular beam epitaxy, as depicted in Fig. 11. In the ultra-high vacuua achieved by H. D. Hagstrum[27] in surface physics research, a molecule-by-molecule deposit of the appropriate elements or compounds, occurs. This happens with suitable accommodation coefficients on a given

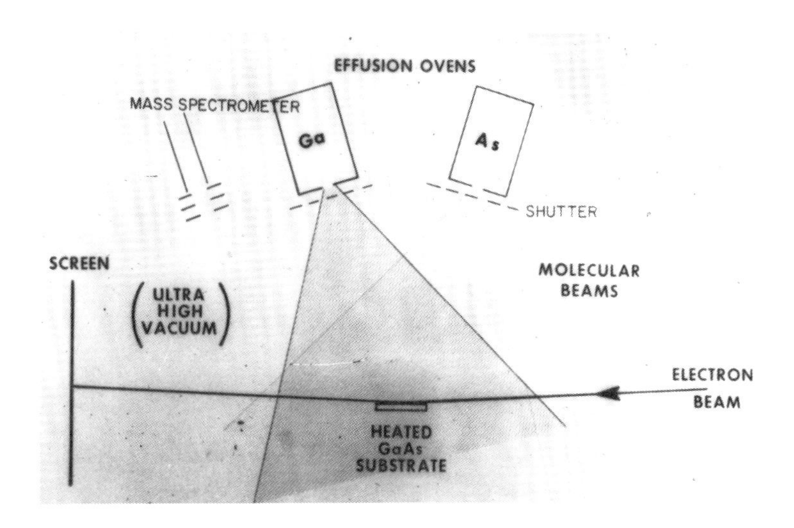

FIGURE 11. BY DEPOSIT OF SELECTED ELEMENTS SUCH AS GALLIUM AND ARSENIC SHOWN IN THE DIAGRAM ONTO CAREFULLY CONTROLLED AND CHARACTERIZED SUBSTRATES, MOLECULAR BEAM EPITAXY HAS ACHIEVED FILM SYNTHESIS ATOM BY ATOM. IN THIS PROCESS LIES THE MATERIALS FUTURE OF MICROELECTRONICS.

substrate, so that truly pristine surfaces never exposed to foreign elements are obtained.[28 29 30]

Similarly, but more violently, ion implantation from fairly high energy nuclear beams is also extensively pursued in the electronics semiconductor industry.[31 32] Many metal alloys of electrical interest have also been made by such bombardment.[33]

For semiconductors, an early preferred process for surface making was an early chemical vapor deposition through silicon tetrachloride on low resistivity silicon base. And further, epitaxial layers could be formed by diffusion. Hence evolved, with further studies, also careful silicon oxide formation control, leading to precise topology of the epitaxial bipolar transistor in the form of silicon-silicon dioxide-metal oxide semiconductor component. This yielded the first field effect surface device.[34] Indeed, new combinations of these materials assemblies were soon forwarded, in which A. Hoerni invented the planar transistor.[35] In this diffusion governed by the pattern of surface oxide from photolithography, combined all the values of oxide masking, photolithography, aluminum metalization, oxide passivation, and of course, epitaxy.

Surfaces for Solar Cells and Other Electronic Units

An interesting forerunner of materials providing these effective structures was the discovery of large area silicon diodes[36] These provided, among other functions, a new level of photocell output (at first about 15 times larger than the best earlier solar energy converters). This solar cell has, of course, furnished the principal power for all earth satellites, and has advanced to a stage of about 20% efficiency, providing a major materials frontier for conventional energy needs throughout the world. But its discovery derives directly from the various semiconductor materials processes and evolution noted. The cell has extraordinary resistance to radiation damage from Van Allen belts or other encounters of the carrier satellite.

In this general context, it is interesting to note that electronic materials development has supported all the operations in outer space from the first earth satellite to now. The first commercial vehicle, Telstar, had, for example, 3600 solar cells, 372 inductors, 1,119 transistors, 64 transformers, 1521 diodes, 5 quartz crystals, 1343 capacitors, 2949 resistors, and 1 yttrium iron garnet limiter. Each of these components indeed represents some phase of materials development noted, or to be noted.

Electronic circuitry continued in the mid-Century to impel materials advances and vice versa, as junction devices dominated systems. Concerning circuit elements, electrolytic capacitors also were brought forward as the needs for electronics components elaborated. By the mid 1950's, it was clear that the intrinsically low impedance of those transistor-related circuits had to be balanced with a higher capacitance, and the tantalum solid electrolytic capacitor, in which the preferred solid (so it wouldn't leak out) was manganese oxide, became a self-healing resource for huge numbers of such circuits. That capacitor had rated voltage of 30 volts or lower, and could be formed, or oxidized, yielding a dielectric constant of about 26, and hence became a highly compact part of the growing culture of film electronics and hybrid circuits. By the mid 1970's, hundreds of millions of these units had been made, of high quality.

The balancing resource of resistors had also a comparable materials evolution. In the pre-electronics times, these needs were generally less than 10^4 ohms per unit, and wire wound resistors could work adequately. Ingenious designs of wrapping held up for some time, in which, for instance, it was learned to make microwires of 0.02" diameter. The required insulation on this wire was itself a materials development of memorable difficulty, involving some of the new Carother's-duPont polyamides, which we oxidized selectively. However, work in the laboratories was intensively begun to use film resistivity, particularly from deposited carbon.

Following earlier trials in Europe, a concentrated program for the development of carbon film resistors by the first systematic CVD (chemical vapor deposition) application for electronic materials formation was undertaken.[37] Related studies indicated the deposited film required more stable substrate than the china rods in which alkali metals could migrate, so a particular alkaline earth porcelain was developed.[38]

These early film components also stimulated interest in many electronic properties of deposited carbon, and were related to the subsequent polymer carbon work in which a combination of graphite and diamond bonding was recognized in solids prepared from densely crossing hydrocarbon polymers.[39] These, and related studies, were the precursors of the present application of chemical vapor decomposition related to varieties of new forms of carbon such as the C_{60} polygon cages and the single-crystal diamond films denoted as the "molecule of the year" for 1990.[40] Indeed, many of the cvd processes functioning in this work use the same ingredients such as methane and hydrogen gas that were applied in the early common resistor film production. The new work has achieved growth rates said to be as much as 1 millimeter per hour of diamond, and many doped films are being studied for semiconductor applications as well.

As noted, the original tactics for contriving electronic materials has meant that components were developed through firm commitment to the underlying science. For instance, in the case of the film capacitors, the dielectric properties of both the films themselves, whether of polymers such as polyethylene terephthalate, or of electrochemically-formed tantalum oxide have been characterized and controlled according to the basic principles of polar and dipolar theory. Indeed, Professor Peter Debye was himself an intimate advisor in the era in which special dipole compounds were synthesized. Some were made to be camphor-like (with rotational freedom in the solid state) and others were halogenated aeromatics. These provided between-film augmentation of the total capacitors themselves. Along with inductors, such capacitors provided both resonant energy storage and filtering of wide-spectrum signals in electronics. In circuit design, the electrical impedance of a capacitor varies inversely with both frequency and actual capacitance. Further, in systems of frequent switching, as demanded by computers and telecommunications, time constants such that the disturbance of on-and-off states can be controlled, demand exceptional stability and reliability of the components. A wide ranging technology has come from this set of functions. For instance, the early use of high-quality, purified paper whose dielectric properties were dominated by the polar hydroxyls, gave a particularly desirable circuit impedance. This stimulated also study of oxidation and hydrolytic protection of the paper. This chemistry is now prominent in the preservation of books and libraries, where most of the world's literacy is threatened by environmental reactions (causing destructive embrittlement) close to or the same as those that were controlled in electronic systems of 1935.[41]

Many other materials have been expertly generated or adapted to the widening demands of electronic circuitry in these years. Silicate mica had many appealing properties for stacked capacitors, and eventually the electrodes were added as silver paste derived from the age-old decoration of china. Truly synthetic capacitor films came later, with polystyrene not having really qualified until the late 1950's, when highly-oriented, high-molecular weight flexible material was developed. However, it had already been transcended by the discovery in the United Kingdom, following W. H. Carothers (duPont) invention of the condensation polymers from polyamides and polyesters. This discovery, of course, was that of polyethylene terephthalate.

The electronics circuit application pushed mylar production beyond the packaging film function that was duPont's initial objective. Still later, polypropylene was also especially adapted, although not, of course, initially developed for foil capacitors. It is interesting, though, that the excellent properties of the mylar polyester reflected exactly the dipoles of polyester solids which were synthesized and investigated by Baker and Yager in 1941-42.

Electronic Materials in Superconductivity and Magnetics

Among the many responses of materials development in the electronics and solid state era, that relating to many body problems of statistical quantum mechanics has been extensive and far reaching. Theoretical and experimental physicists have worked together with dramatic extensions of the horizons of science, on which new and exceptional materials have been prominent features. The realm of superconductivity has been a most visible and popular arena for this activity, reflecting, of course, the very reasons for the fundamental interest in conductors and insulators, which were cited earlier. This stage began when Kamerlingh Onnes at Leiden in 1911 found that at 4.2°K, the voltage of a wire of frozen mercury dropped to 0, as the current continued to flow. Thus was opened a vision of electromagnetism and electricity, which still challenges understanding and tantalizes exploitation. Although scores of materials were explored in seeking higher temperatures for superconductors, there were few new or encouraging directions before the work of Hardy and J. K. Hulm at the University of Chicago in 1952, who showed systematic evidence of the transition to zero resistance in intermetallic compounds of silicon or germanium with transition metals, especially when these existed in a beta tungsten structure. Indeed, V_3Si had a critical temperature T_c for superconductivity of 17°K. Matthias and coworkers soon extended this domain to related materials, especially niobium3 tin (Nb_3Sn) to 18°K[42] The ideas of chemical "average valence" related to T_c were then extensively applied throughout the Periodic Table, and in the 1959-72 time, total number of known superconductors was doubled over the total world listing since 1911.[43] This chronicle, being extensively enlarged right up to the present day, provides a fascinating vision of how materials science and engineering can support the elegant explorations of modern physics. Although the historic findings recently (1986) of much higher temperature superconducting ceramics $(LaBa)_2Cu0_4$, by K. A. Muller and J. G. Bednorz challenge new usage, the materials development with

Nb_3Sn and its derivatives marked the modern period of superconductor applications, with T_c of 23°K for Nb_3Ge[44]. These compounds sustain magnetic fields above 88 kilogauss, and indeed utility at 150 kiloersteds has enabled extensive service of wire-wound magnets at temperatures below 20°K. These are used in national laboratories for particle accelerators, and all over the world for nuclear magnetic resonance imaging in physiology and medicine[44] In Figure 12 are shown some of the magnetic field tolerances of a range of superconducting compounds, with transition critical temperature shown on the abscissa.

In electronics materials development, an equivalent role to the electricity we have illustrated comes from magnetic function. Although metallurgy has a vast inventory of

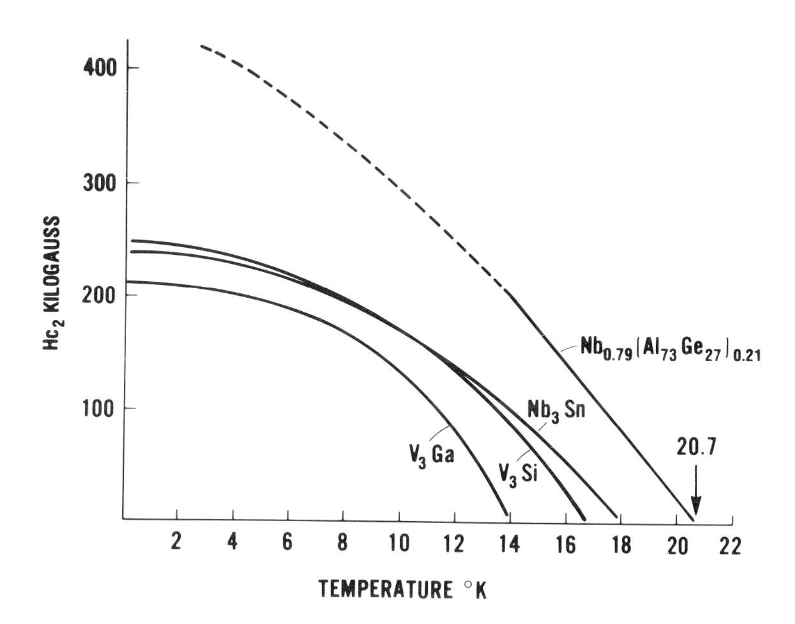

FIGURE 12. PRESENT TECHNICAL AND INDUSTRIAL USE OF SUPERCONDUCTORS IS WITH WIRES AND CABLES OF COMPOSITIONS SHOWN ABOVE, WHICH PROVIDE AT TEMPERATURES BELOW 20°K IN HIGH MAGNETIC FIELDS FACILITIES FOR MAGNETIC RESONANCE IMAGING (MRI) IN MEDICINE AND BIOLOGY AND HIGH FIELD MAGNETS FOR MANY NUCLEAR PARTICLE EXPERIMENTS.

magnetic materials, the features of electronics circuitry repeatedly noted make the discovery of nonmetallic high resistivity magnetics crucial. This universe was achieved through the ferrites, especially in the work of the Philips Research Laboratories in the Netherlands during the early 1940's[45] The resistivities of these compounds can range from 10^2 to 10^6 ohm centimeters, in contrast to the quantities we scaled earlier for ferromagnetic metals of about 10^{-5} ohm cm. Chemistry of the ferrites, generically expressed as MFe_2O_4, where M is a divalent ion of magnesium, zinc, copper, nickel, iron, cobalt, manganese, or mixtures of those, is a major sector of the evolution of electronic materials. Accordingly, the fundamental analysis of behavior by Louis Neel at the University of Grenoble in France has conferred yet another set of guiding principles which have subtly and effectively stimulated vast materials generation.[46] Indeed, the complementary qualities of magnetic materials for metals and alloys and those of the insulating ferrites themselves provide brilliant chapters in electronic materials developments.

Among the metals, for instance, the discovery of cobalt/rare earth permanent magnets founded new magnetic technologies providing product of coersive force H_c and residual induction, B_r to yield maximum energy product (BH_m) of 30 x 10^6, oersteds -- far beyond any magnet strengths up to their time.[47]

Then among the ferrites yielding such new electronics as magnetic 'bubble' memories, a reawakened realm of inorganic chemistry responded to the physical insights about orthoferrites, which were needed in single crystal form. The demanding physics and electromagnetics of these devices led to the discovery of the rare-earth garnets as effective media. These required delicately balanced electromagnetic and crystallographic qualities within the single crystal garnets, optimized for magnetic properties. Growth of these crystals was frequently done in lead oxide-lead fluoride-boron oxide fluxes. It represented an expert balancing of inorganic components. Figure 13 depicts one of the gemlike crystals with large and technologically useful faces of the general type whose chemical composition is shown in Figure 14.[48] The large unit cells and complex composition such as in the two examples shown containing two or three rare earths as well as calcium, germanium and iron demonstrate graphically the versatility of modern electronic materials synthesized to fit the exacting parameters of circuitry. It is interesting that in the work on these garnets and their absorption spectroscopy, it was found that the neodymium-doped yttrium aluminum garnet had a low energy pump band so that a high-power laser action could be obtained from about a 360 watt input. These Nd:YAG systems have become one of the most highly applied laser sources, ranging from industrial processing to smart weapons target seeking.

New Directions of Materials Development

From our own experience, we should expect the arena of materials for electronics to enlarge steadily in the future. The advent of photonics to which we have been committed following the discovery and development of the laser has already engendered the extension of solid state electronics through opto-electronics circuitry and systems. The

FIGURE 13. LARGE SINGLE CRYSTALS OF RARE EARTH GARNETS SHOWN ABOVE PROVIDE A NEW REALM OF MAGNETISM AS WELL AS OF MEDIA FOR LASER ACTION.

	$Y_{1.88}Lu_{.2}Ca_{.92}Ge_{.92}Fe_{4.08}O_{12}$	$Y_{1.64}Lu_{.3}Eu_{.1}Ca_{.96}Ge_{.96}Fe_{4.04}O_{12}$
Strip Width d (μm)	6.2	6.6
$4\text{-}M_S$ (Gauss)	171	140
Curie Temp. (°K)	465	460
Mobility (cm/sec-Oe)	2000	1400
Q $\left(\frac{\text{Anisotropy Field}}{4\pi M}\right)$	4.7	7.0

FIGURE 14. THE COMPLEX UNIT CELLS OF THESE GARNETS CAN BE ADJUSTED CHEMICALLY TO MEET EXACTING MAGNETIC DOMAIN DEMANDS SUCH AS IN "BUBBLE" MEMORIES. THE SENSITIVITY TO CHEMICAL COMPOSITION APPEARS ABOVE, WHERE PARAMETERS FOR MAGNETIC BEHAVIOR ARE OPTIMIZED BY THE COMPOSITIONAL FORMULAS SHOWN.

course we have used in forming this new frontier is closely linked to the experiences of semiconductor and other electronics materials development described before, especially for the element silicon. A benign consequence of the transistor and integrated circuits era, based on ultra pure silicon chemistry and technology, shows up in our transfer of this knowledge to the preparation of silicon dioxide for glass fiber lightguides. This, too, is even suitably doped with Ge or other things to control the reflection and mode conversion properties of the optical fiber.

These photonic media are three orders of magnitude more loss-free than any glass has ever before been shown to be (at about 10^9 higher frequency than the conventional electronic circuit). Accordingly, a dielectric surpassing even polyethylene and other polymers, as well as quartz, (synthesized by hydrothermal crystallization) has vastly (and globally) enriched propagation of waves and signals.[49]

As usual, the electronics/photonics materials chemistry has also provided useful mechanical improvements. In meeting the needs for cable handling, methods of synthesizing (by chemical vapor deposition, CVD) and of immediately coating the glass fibers by in situ polymerization have yielded median (and uniform) tensile strengths of 5.25 GN/m^2 (750,000 lbs./$in.^2$.) This demonstrates a new scale of possibility (with simplified preparation) for fiber glass in structural composites.

Correspondingly, the connections with still other electronic materials science and engineering are surging ahead. Need for new connections between using electrons and photons was realized in beginning our research and development programs. Photons have comparatively high energy of about 1 electron volt, in comparison to the electromagnetic field of modern electronics, and have effective dimensions of about 1 micrometer. So the circuit quality effected by materials requires new versatility. One important aspect is how electrooptic materials can respond, shown largely in their refractive index change, Δn_i, caused by electronic inputs. This influence of the electromagnetic field E_j relates to the refractive index (square root of the dielectric constant) noted in the beginning in the form $\Delta n_i = 1/2\ n^3_i r_{ij}$ (the electro-optic tensor coefficient) x E_j the applied electromagnetic field. Materials developed for their electronic circuit qualities as filters and other functions, but that are of increasing importance in this electro-optic frontier, include a whole class of ferroelectrics. Examples are lithium niobate as well as strontium, barium and niobium oxides, barium titanate and potassium dihydrogen phosphate. Likewise, in still other analogy to the electronic functions, the semiconductors gallium arsenide, indium phosphide, and cadmium telluride show interesting figures of merit, as do a host of organic solids such as nitroaniline and other aniline derivatives. Chemical and material features of these compositions challenge again the chemical synthesis and aggregation of polymers as well as a host of more conventional solids.[50] Accordingly, it is heartening that the work of decades ago in electron conductivity as well as insulating properties of variety of macro-molecular solids is now being extended into exciting novel electronics/photonics capabilities.

Evidently, the chemistry and physics and engineering of materials have evolved vigorously, especially in the last half century, in concert with the scientific and technical emergence of a communication and information age. It seems that this interaction has also markedly enhanced the interplay of industrial technology and academic research and development. Likewise, both in educational and industrial as well as governmental enterprises, the electronics/materials initiative has reduced traditional disciplinary separations and barriers. Accordingly, it seems that models are manifest for similar advances with materials in the energy, biomedical, construction, and various other areas of socioeconomic urgency. Throughout, close connections have been kept with all phases of chemistry. Indeed, the manufacture and maintenance of many electronic systems increasingly resemble sophisticated chemical synthesis. The steady evolution of self-organizing matter, especially in the spread of nanosystems, promises dramatic advances in photonic and electronic facilities, as well as in such new areas as the micromachines discovered by Gabriel and his colleagues,[51] academic and industrial centers. These mechanisms, directly derived from the thin film lithographic and chemical processes for film technology, may, of course, be ultimate media for joining electronic information with physical action. And, indeed, as the next stages of history take shape, concepts of matter and electricity, contributed by Faraday, Maxwell, Thomson, and Wilson, among others, in the national environ of the Royal Society of Chemistry will already have shown the way.

1. Perrin, M. W. *Research, 6,* Chem. Ind. (London) 396 (1955).
2. Yager, W. A., and Baker, W. O., *J. Am. Chem. Soc., 64,* 2164 (1942); Baker, W. O. and Yager, W. A., *Ibid., 64,* 2177 (1942).
3. "Macromolecules", Bovey, F. A. and Winslow, F. H. Ed.s, P.17 ff, Academic Press, N.Y., 1979.
4. Baker, W. O., in "Advancing Fronts in Chemistry", Twiss, Sumner B., Ed. pp. 105-155, Reinhold Pub. Corp. N.Y., 1945.
5. Hopkins, J. L., Baker, W. O., and Howard, J. B., J. Appl. Phys, 21, 206(1950).
6. Baker, W. O., VII and authors,ff, in "Advanced Technology", Abelson, P. H. and Dorfman, M., Eds., Washington, D.C., Am. Assoc. Adv. of Science, 1980.
7. Gilmer, G. H., J. Crystal Growth, 42 3(19977); Science, 208, 355 (1980); Jackson, K. A., J. Cryst. Growth, 24, 130(1974).
8. Pfann, W. G. Trans. AIME, 194, 747, (1952); U.S. Pat. 2,739,008.
9. Pfann, W. G. et al, "Phys. Rev.," 94, 489 (1953). Patel, J. R. & Chandhuri, A. R., J. Appl. Phys. 34, 2788 (1963); Phys. Rev., 143, 601 (1966). For further review of this work, see Baker, W. O., Jour. of Materials, 2, 915, (1967), including Theurer, H. C., Jour. Metals, 8, 1316(1954).
10. Rend, W. J. Jr., "Dislocations in Crystals", New York, McGraw Hill, 1953; Cottrell, "Dislocations and Plastic Flow in Crystals; London, Oxford Univ. Press, 1953.
11. Vogel, F. L. Jr., Pfann, W. G. Corey, H. E. Thomas, E. E., Phys. Rev., 90, 489(1953).
12. Compton, K. G., Mendizza, A., and Arnold, S. M., Corrosion, 7, 327 (1951).
13. Herring, C. and Galt, J. K., Phys. Rev., 85, 1060(1952)
14. Greiner, E. S., Gutowski, J. A., and Ellis, W. C., J. Appl. Phys., 2489(1961).
15. Mott, N. F., "Metal-Insulator Transitions", London, Taylor and Francis, 1974.
16. McWhan, D. B. Rice, T. M., Remeika, J. P., Phys. Rev. Letters, 23, 1384(1969).

17. Wagner, C. and Schottky, K. Z. Phys. Chem., B11, 163(1930).
18. Bloch, F. Z. Physik, 52, 555(1928).
19. Wilson, A. H., Proc. Roy. Soc. London, 133, 458(1931).
20. Wilson, A. H., "Semiconductors and Metals", Cambridge Press, 1939. London, 133, 458(1931).
21. Bardeen, J. and Brattain, W. H. Phys. Rev., 74, 230(1948).
22. Shockley, W., Bell Sys. Tech. Journ. 28, 435(1949).
23. "Semiconductors", ACS Monograph 140, Hannay, N. B., Ed, Reinhold, W. Y., 1959, See particularly Lander, J. J., Chap. 2; Tanenbaum, M., Chap. 3; Fuller, C. S., Chaps. 5 and 6; Hobstetter, J. N., Chap. 12; Hutson, A. R., Chap. 13 and Garrett, C. G. B., Chap. 15.
24. Sauer, H. A. and Flaaschen, S. S., Proc. Seventh Electronic Components Symp.. Washington, D.C., p. 41 (1956); J. Am. Ceramic Soc., 43, 297 (1960).
25. Arthur, J. R., Jr. J. Appl. Phys., 39, 4032 (1968). Christenson, H. and Teal, G. K., U.S. Patent 2,692,839 filed April 7, 1951; issued 1954.
26. Theurer, H. C., Lleimack, J. J. , Lozr, H. H., and Christensen, H., Proc. IRE, 46, 1462 (1960).
27. Hagstrum, H. D., Science, 178, 275 (1972).
28. Cho, A. Y., J. Vac. Sci. Technol., 8, 531 (1971).
29. "The Technology and Phys. of Molecular Beam Epitaxy," Parker, E. H. C., Ed., New York, Plenum Press, 1985.
30. Panish, M. B., Quiesser, H. J., Denich, L., and Sumski, S., Solid State Electronics, 9, 311 (1966).
31. Ohl, R. S., Bell Sys. Tech. Jour., 31, 104 (1952), U.S. Pat. 2,750,541, filed 1950, issued 1956.
32. McRae, A. U., p. 329 in "Ion Implantation in Semiconductors," Ruge, J., and Graul, J., Banlin, Springer Verleg, 1971.
33. Poate, J. M., DeBonte, W. J., Augustyniask, W. M., and Borders, J. A., Appl. Phys. Letters, 25, 698 (1974). Also see Ion Implantation Metallurgy, ed. Priece, C. M. and Hirvonen, J. K., N.Y., AIME (1980).
34. Atalla, M. M., U.S. Patent 3,206,670, 1965, filed March, 1960, Kahng, D., U.S. Patent 3,102,230, 1963, filed 1960, Kahng, D. and Atalla, M. M., paper at Solid State Device Research Conf., Pittsburgh, Pa., June, 1960.)
35. Hoerni, J. A., IRE Trans., Trans. of Electron Devices, Vol. 8, pg. 178, 1961, U.S. Patent 3,025,589, 1962, filed May, 1959.)
36. Pearson, G. L. and Fuller, C. S., Proc. of the IRE, 42, 760, (195); Chapin, D. M., Fuller, C. S., Pearson, G. L., Jl. of Applied Physics, 25, 676, (1954).
37. Grisdale, R. O., Pfister, A. C., and VanRoosbroeck, W. Bell System Technical Journal, 30, 271, (1951.)
38. Rigterink, M.D., Bell Labs Record, 25, 464 (1947).
39. Baker, W. O., Winslow, F. H., Proc. of 12th International Congress of Pure and Applied Chemistry, Sept. 1951; Winslow, F. H., Baker, W. O., Pape, N. R. and Matreyek, W., Jl. of Polymer Science, 16, 101, (1955); Baker, W. O. and Grisdale, R. O., U.S. Patent 2,697,136, 1954.)
40. Science, 250, 1640, 1990.)
41. McLean, D.A., Industrial Eng. Chem., 39, 1457, (1947); Sauer, H. A., McLean, D.A., and Egerton, L., ibid., 44, 135 (1952).
42. Matthias, B. T., Gaballe, T. H., Keller, S., and Corenewit, E., Physical Review, 95, 1435, (1954).
43. Matthias, B. T., Phys. Rev., 97, 74, (1955); Roberts, B. W., Jl. of Phys. Chem. Ref. Data, 5, 581, (1976).
44. Kunzler, J. E., Buehler, E., Hsu, F.S.L., and Wernick, J. H., Phys. Rev. Letters, 6, 89, (1961); Hulm, J. K., Kunzler, J. E., and Matthias, B. T., Phys. Today, 34, 34, (1981).
45. Snoek, J. L., New Developments in Ferromagnetic Materials, N.Y., Elsevier, 1947; Also see Takei, T., Ferrites, Proc. Int'l Conf., 436, Hoshino, Y., Iida,S. and Sugimoto, M., Eds., Baltimore Univ. Park Press, 1971.
46. Neel, L., Ann. Phys., 3, 137, (1948).

47. Nesbitt, E. A., Wernick, J. H., and Corenzwit, E., Jl. of Appl. Phys., 30, 365, (1959); Nesbitt, E. A., Willens, R. H., Sherwood, R. C., Buehler, E., and Wernick, J. H., Appl. Phys. Letters 12, 361, (1968).
48. Bobeck, A. H., VanUitert, L. G., etal, Appl. Phys. Letters, 17, 131, (1970).
49. For brief review, see Baker, W. O., Metallurg. Transactions A, 8A, 1205 (1977).
50. Glass, A. M., MRS Bulletin, 13, xii and 14 ff (1988) and other authors.
50. Baker, W.O. et al, "Advancing Materials Research", Psaras, P.A. and Langford, H.D. Eds., Wash., D.C., Nat'l Acad. Press, 1987.
51. Gabriel, KJ., Trimmer, W.S.N., and Mehregany. Tech. Digest, "The 4th Int'l Conf. on Solid-State Sensors and Actuators", IEE Japan, pp. 849-852 (1987).

Development of Liquid Crystal Materials for Information Technology

G.W. Gray

MERCK LTD., WEST QUAY ROAD, POOLE, DORSET, BH15 1HX, UK

1 INTRODUCTION

The section of the 150th Anniversary Annual Chemical Congress in which this paper is presented has as its overall theme "Milestones in 150 Years of the Chemical Industry." The sub-section in which it is located is Information.

This is highly appropriate, as the Milestone about which I will speak concerns the discovery/invention of a new and valuable family of liquid crystal materials designed to meet the requirements of an excellent liquid crystal display mode that was discovered at about the same time, both within an electronics infrastructure well able to cope with the progression of the two inventions through technical aspects, packaging and into the market place. It must be emphasised therefore that part of the success of the new liquid crystal (LC) materials has been due to the timeliness of their discovery - there existed a known device need for them and conditions existed for their exploitation. This distinguishes inventions in the Materials Science field from those in other areas of science. Inventions of chemical reactions, physical techniques or theories advance science whenever they occur, but inventions of new materials must, to be most successful, occur within the right time frame such that other areas of science and technology are equipped on a knowledge base to receive them and use them to maximum advantage.

The story I have to tell also reflects the fact that, today, few real advances in science and technology can be credited to a single individual. In the case under discussion, we will see that the knowledge base of organic chemists at the University of Hull in structure/property relations for LC systems was used to develop and produce materials tailored to the requirements of a newly proposed mode for displaying information through an LC electro-optical effect. This and the optimisation of the new materials in eutectic mixtures involved physicists, followed by industrial chemists to produce the materials in large quantities, and ultimately the skills of electronic engineers and applied physicists to produce quality displays for the market place. The success has therefore developed through a highly collaborative situation involving organic chemists in a University (University of Hull) and in industry (BDH Ltd), physicists in a Government Research Establishment (RSRE, Malvern) and applied physicists/electronic engineers in the many electronic companies wherein matters were finally driven forward through the technical aspects of device design and on into the marketplace.

The original materials invented at Hull are commonly known as cyanobiphenyls - we will focus on precise structures presently - and are now referred to simply as CBs. They were invented at a time when physicists were becoming aware - late 1960s - that liquid crystals probably might have a role to play in the area of devices for the attractive, low power consumption display of information. Progress was however frustrated by deficiencies of stability, colour etc of existing LC materials, and the discovery of the CBs and the first publication[1] of their properties was a very significant event. F.Funada of the Sharp Corporation said[2] in the RSC Symposium on Fine Chemicals for the Electronics Industry II in York in 1990 -

> "These were the first materials which had good chemical stability, non-optical absorption within the visible light range, a practical nematic (liquid crystal) range and good electro-optical characteristics, and (therefore) these materials should be memorable in the history of liquid crystal displays."

- great praise from a centre of excellence.

Put another way, the CBs provided the first LC materials which could be used for the production of stable long-life, high quality LC displays operating on the twisted nematic mode, and as such, these materials provided the firm foundations upon which today's highly successful Liquid Crystal Display Industry is based.

Sadly, situations expressed in dollars make more

impact than words. Currently, overall sales of LC displays run at about 1 billion dollars per annum. Now, use of LC displays has moved on from simple watch and calculator displays into the large area devices used for personal and desk-top computers and television. Again using data from Funada[2] --

Personal computers made in	1989 --	about 20 million
Predicted growth expected in	1995 --	to ca 40 million

Some 60% of these, i.e., 24 million sets, are predicted to be of the portable, LC type. Limiting thoughts to just PCs, sales of flat panel displays are forecast to be about 7 billion dollars in 1995.

It is also predicted that the lines representing the rising sales of both CRT displays and LC displays will cross in 1995, with sales of LC displays EXCEEDING those of the CRT type. This gives some idea of the current and forecast states of the LC Display Industry which was first triggered off by the discovery of the CBs and sustained by the quality of the materials and other related materials developed subsequently on the basis of the CB structural model and as competitors to them in the market.

2 HISTORY

So how[3] did all this come about, and who and where were the main players?

In 1967, John Stonehouse, Minister of State for Technology, visited the Royal Radar Establishment at Malvern and had discussions with Dr - later Sir George Macfarlane. Through that visit the seminal idea developed with the Minister that a solid state alternative to the shadow mask tube was highly desirable, and this was later put to David Parkinson and Cyril Hilsum of RRE. By 1968, responsibility for displays was in the hands of Dr Hilsum, and together with Leslie Large, and as a first move, a meeting on liquid crystals was set up in London to explore whether these (to most people strange) materials called liquid crystals could possibly assist in realising the solid state alternative in question. Several presentations were made at the meeting, and as a consequence, it was decided that the "the Hull man" - George Gray - should be put on a contract - this despite the fact that one participant had said " Liquid Crystals may have a minor role for displays in high ambient lighting, but they will make no impact on Black and White or Colour TV."

So work commenced in 1970 with the author and one post-doctoral Fellow, and later a technician, at the University of Hull.

In the meantime, work had been going on at the David Sarnoff Research Centre of RCA at Princeton, New Jersey on some possibilities for display applications of liquid crystals[4]. Clearly, any applications of liquid crystals depend upon the appropriate liquid crystal phase being thermodynamically stable at ambient temperatures. In 1969, the only real candidates for this category of material were Schiff's Bases such as 4-methoxybenzylidene-4'-n-butylaniline (MBBA)[5] which melts

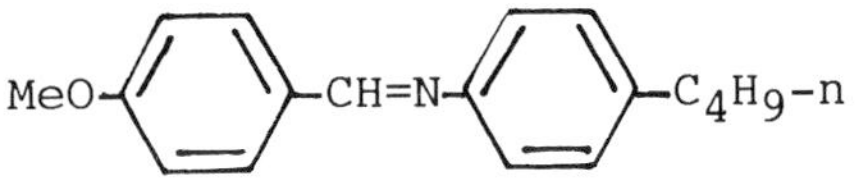

MeO–C₆H₄–CH=N–C₆H₄–C_4H_9-n (MBBA)

at 22°C, forms a nematic LC phase which is fluid but ordered and in which the rod-like molecules are statistically parallel to one another, but there is no regularity of ordering of the molecular centres or ends, i.e., a liquid crystal state - which persists until the isotropic phase is formed at 47°C

C 22°C N 47°C Iso

Such a material and mixtures of it with appropriate homologues of MBBA provided workable, room temperature nematic phases of negative dielectric anisotropy

$$\Delta\varepsilon = \varepsilon_{\parallel} - \varepsilon_{\perp} = \text{a negative value}$$

Thin films of such materials can be aligned homogeneously - i.e., with the nematic director (the average direction of orientation of the long axes of the rod-like molecules) parallel to the glass surfaces and are optically clear.

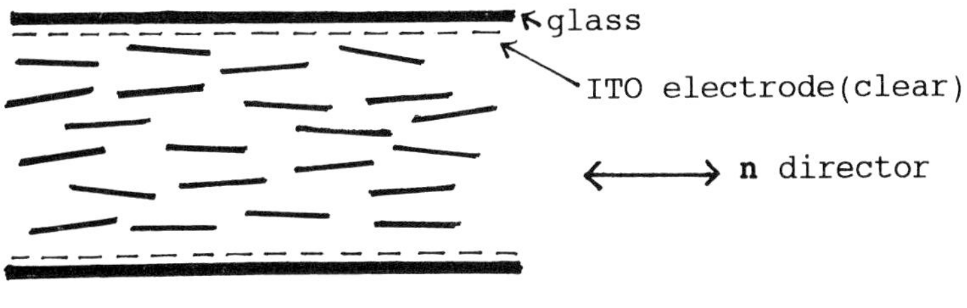

glass

ITO electrode(clear)

n director

Often, the Schiff's bases were sufficiently impure to be conducting, but if not were doped with ionic dopants. In an anisotropic system such as this, the conductivity is anisotropic, and putting volts across the film by means of the ITO electrodes produced undulation modes in the aligned phase which, with increasing voltage, developed into an electrohydrodynamic turbulence -- a stirring effect which rendered the film, or those parts of the film that were addressed electrically, turbid and scattering. Bright scattering regions against a clear background could be produced - i.e., information in the form of numbers, characters etc - could be produced through an electrically controlled optical effect. Due to the hydrolytic instability of Schiff's bases, and difficulties in handling them and maintaining them pure, devices had to be constructed with great care to exclude moisture. Lifetimes were therefore poor as moisture permeated edge seals and filling holes, but by their very existence, these displays stimulated research activity in the field.

3 EARLY RESEARCH AND THE DEVICE SCENARIO

In the early part of the Hull programme, we were therefore seeking superior, more stable, room temperature LC materials of negative dielectric anisotropy. With funding of £2177 per annum for all costs, we examined stilbenes, chlorostilbenes, and a range of esters including heterocyclic systems, all with rather depressing results. This was the position until the news broke of a new display mode originating from work by Fergason[7] and Schadt and Helfrich[8] which was eventually published in 1971 - 1973.

This device, whose mode of operation is in Figure 1 and summarised in the text below, required a room temperature nematic of positive dielectric anisotropy.
As shown, the unique aspect of the twisted nematic cell is that in the OFF state, the nematic director is rotated through 90° by the mechanical influence of two supporting glass plates treated to give the homogeneous alignment at 90° with respect to one another. As a result of the quarter helix so generated, plane polarised light entering the cell (thickness 6 - 15 μm) from above with its electric vector parallel to the alignment direction, is transmitted through the cell, with guiding of the plane of polarisation through 90°, and emerges from the back plate with its plane of polarisation such that it passes through the back polariser (set crossed with respect to the front polariser). The OFF state cell is therefore optically clear. With a nematic of positive dielectric anisotropy,

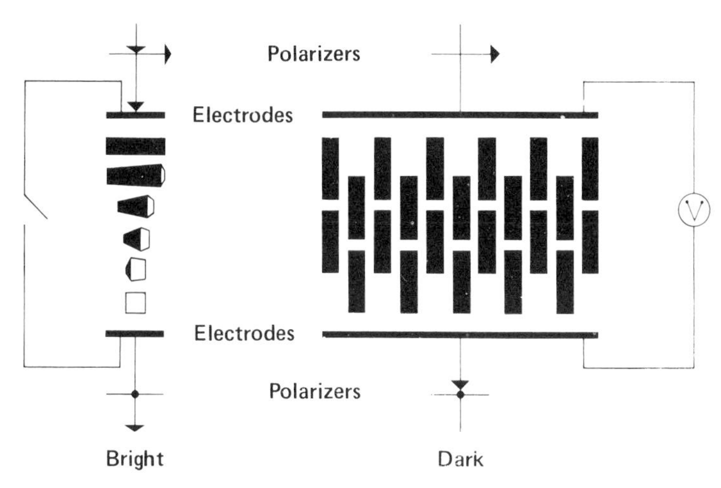

Figure 1 The twisted nematic liquid crystal cell

the applied field converts the twisted film into a homeotropically aligned film with the molecular long axes perpendicular to the cell surfaces. The ON state is no longer light guiding, the light cannot be transmitted and the cell or any addressed area is extinct (black) - also, of course, with a back plane mirror, the cell is bright (OFF) and black (ON) in reflection.

But even in the ON state, the original homogeneous, planar alignment is retained in a very thin boundary layer at each surface. On switching OFF, the nematic therefore relaxes back under the aligning influence of these boundary layers to regenerate the light guiding quarter helix. Black addressed areas then become bright again.

This has proved to be an excellent display system. Contrast is high (black/bright) and the display itself is passive, involving just a field effect (not conductivity). High resistivity materials can therefore be used and are essential. The device operates with ambient light at low voltages (battery operation), with minimal power consumption, and cells can be edge-lit for dark operation. Setting the polarisers parallel gives bright data on a black background, and use of one colour polariser gives coloured data on a clear background or vice versa.

Due to the quality of the device and its appeal to customers, pressure quickly developed for a progression from simple direct drive displays operating on a seven segment, figure of eight bar electrode format into multiplex addressed dot matrix displays capable of portraying much more complex information and ultimately into flat panel displays, capable of full colour and in which each picture element (pixel) is addressed by its own thin film transistor - i.e., colour TV operation.

Even today, well over 90% of LC displays marketed operate on the twisted nematic mode. In the next few years, this situation will certainly change as the supertwisted nematic display (STN)[9] and electrically controlled birefringence (ECB) displays[10] command a bigger share of the market place, but the former of these is still related to the original device of Fergason and Schadt and Helfrich, involving a larger helical twist (up to 270°) engendered by a chiral dopant, put possessing the great advantage of bistabilty, such that data can be stored in the OFF state and recalled.

4 THE CYANOBIPHENYLS

The change in need from a material of negative to one of positive dielectric anisotropy radically altered the organic synthesis situation. The destabilising, inter-ring, centre linkage of Schiff's bases like MBBA, of stilbenes and of azoxy compounds could now be discarded and a direct ring-ring linkage used, in the knowledge that a terminal cyano- group (giving a strong positive dielectric anisotropy) could be used and still guarantee that the shorter molecules would be capable of exhibiting LC properties. This knowledge had derived from much earlier fundamental studies of structure property relations for LC materials, studies in which it had been clearly demonstrated that of many terminal groups, the cyano- group stood strongly at the head as the one promoting nematic properties to the greatest extent and giving high nematic-isotropic transition temperatures. The conceived structure was therefore a 4-cyanobiphenyl:

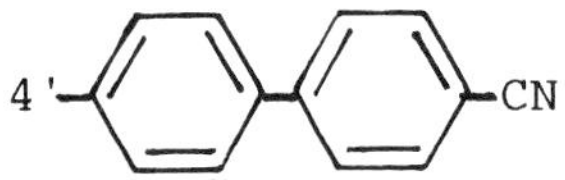

4' CN

At the 4'-position, it now seemed sensible to locate a terminal alkyl or alkoxy group -- desirable end functions in many LC systems. Moreover, the one doubtful and unpredictable parameter was the melting point of such materials. If this were too high, the LC properties would be monotropic, achieved only on cooling cycles if the crystallisation process were subject to supercooling. Use of terminal alkyl or alkoxy groups would allow a whole range of 4'-substituted 4-cyanobiphenyls to be surveyed[11] with from 1 to 18 carbon atoms in the end groups, improving the odds that at least one homologue would be a good, low melting material, so making a room temperature nematic phase available that would be thermally, chemically, and photochemically stable, colourless, of low viscosity, and strong positive $\Delta\varepsilon$.

From the plot of the transition temperatures for the 4'-alkyl-4-cyanobiphenyls shown in Figure 2, it is seen that three materials, the pentyl,hexyl and heptyl homologues had indeed low melting points (C-N), and two, because of the alternating T_{N-I} values, had good nematic phases ranges up to 30-40°C. Significantly too, the octyl homologue gave a low melting smectic A material. All four

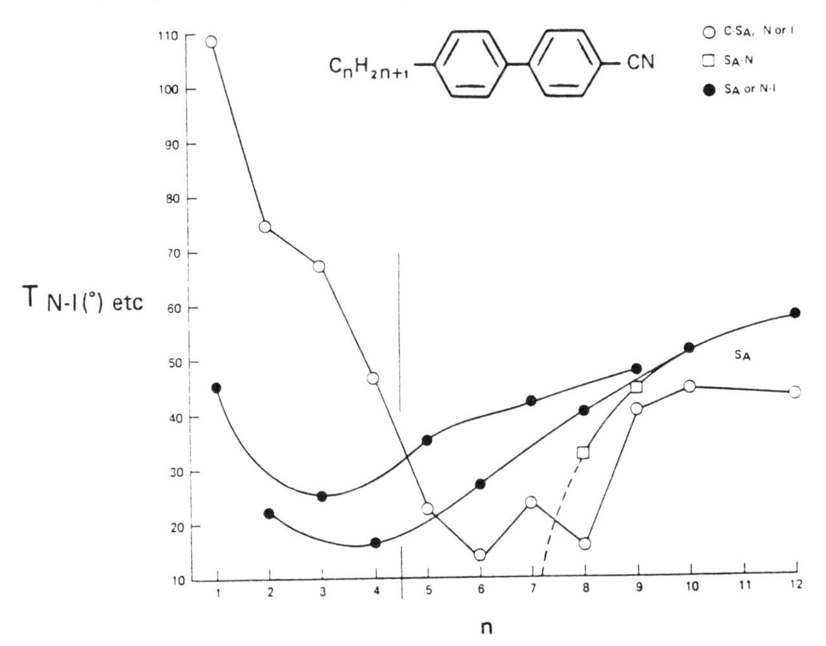

Figure 2 Mesophase transition temperatures plotted against number of carbon atoms in the terminal alkyl chain for the 4'-alkyl-4-cyanobiphenyls

materials supercool in the N or S_A phases for very long periods at temperatures well below ambient, and it is a point of considerable significance that not only did these materials provide nematic compounds of immense importance to the LCD industry, but also materials for smectic displays developed later[12], and more importantly still, a range of single component, room temperature liquid crystal materials for fundamental studies of a great range of physical parameters whereby the theory of anisotropic LC states has been immeasurably advanced and strengthened.

The alkoxy compounds of course melted higher, but also had higher T_{N-I} values. This was valuable, because individual alkyl analogues or mixtures of them could not provide high enough T_{N-I} values for realistic device applications. Eutectic mixtures of the 4'-alkyl (K series) and 4'-alkoxy (M series) materials did however give systems with improved phase ranges. Work done at RSRE, Malvern soon led to mixtures of commercial interest, and BDH Ltd was approached to do large scale synthesis and marketing of materials. This the company did on a remarkably short time scale. Nematic mixture E4 was on sale late in 1973; remember that the first cyanobiphenyl was made in the University laboratories only in mid-1972! E4 was nematic from its m.p. of -4°C (crystallisation temperature $\ll$-4°C) until 61°C. However, for a real commercial breakthrough, a m.p. of -10°C was needed and a T_{N-I} of about 60°C.

The Hull University group therefore developed[13] the 4"-alkyl-4-cyano-*p*-terphenyls to achieve mixture components of very high T_{N-I} which would enable these criteria to be met. 4"-Pentyl-4-cyano-*p*-terphenyl was a most useful material, C 130°C N 239°C Iso, and its incorporation into mixtures gave development of E7, C -9°C N 59°C Iso. BDH Ltd was selling E7 by late 1974. From that time, commercial success was assured and the mixture E7 and many others developed later enabled quality nematic displays to be put on the market by electronic companies for the first time.

Since then, the situation has not remained static. The quality of the devices led to their large scale manufacture, mainly in Japan, the Far East and the USA, engendering demands from customers for similar, but more complex displays that could only be produced by multiplex addressing,[14] and also producing some device fabrication problems that had not been foreseen.

One fabrication problem was the occurrence of ugly discontinuities in devices arising from reversed twist areas. These arise on switching OFF, when some areas change from the homeotropic state to the twisted state by forming a right-handed and others a left-handed quarter helix. This was overcome by preparing chiral CBs - see below - such as S-4'-(2-methylbutyl)-4-cyanobiphenyl (n = 1) and S-4'-(3-methylpentyl)-4-cyanobiphenyl (n = 2).

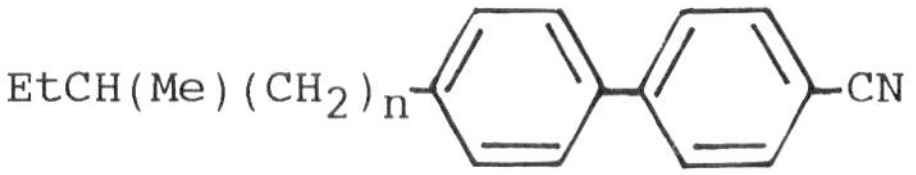

$$EtCH(Me)(CH_2)_n-C_6H_4-C_6H_4-CN$$

Very small amounts of such materials gave the nematic host a tweak on switch OFF to give wholly either a right- or a left-handed quarter helix. Again there was an additional fundamental knowledge spin-off. When n = 1, the compound is a right twister and when n = 2, it is a left twister, and the SED/SOL twist sense rules could be formulated.[15]

Multiplexing of devices also raised problems, as the threshold sharpness of the early E mixtures was not good enough. This was solved by adding to the mixtures components which did not contain terminal cyano-groups, e.g., 4'-(2-(*trans*-4-alkylcyclohexyl)ethyl)-4-alkyl-2-fluorobiphenyls. Although these are not strongly positive in dielectric anisotropy, good threshold sharpness curves could be achieved, and it was eventually understood[16] that this was possible because the non-cyano- additives were minimising loose, antiparallel pairwise correlations between terminally cyano-substituted molecules and so were altering critical physical parameters such as elastic constants of the mixtures.

The foregoing will have made it clear that the discovery of the biphenyls/terphenyls met a perceived need for an excellent display mode (the twisted nematic mode) and gave the embryo liquid crystal display industry the confidence to grow and develop through quality products into the major industry which it now represents. It has also been stressed that there has been great intellectual benefit from the stimulus provided by these materials for reliable physical studies of the liquid crystal state and consequent developments of theory. The simple concept of removing the functional groups used to link up ring systems in earlier mesogens and replacing these groups by direct bonds has therefore been very significant. This principle, used by the Hull group with BDH Ltd to give the biphenyls/terphenyls, has also been seized upon extensively by others to give a wider range of new materials with excellent LC properties. This has been beneficial to the chemical industry and to the development of fundamental knowledge.

Thus,as shown in the scheme on the following page which sets out a historical sequence of nematogen development, E Merck in Darmstadt quite quickly (1976-77) developed[17-18] the PCH, BICH and CCH analogues of the biphenyls and terphenyls, wherein one or more of the phenyl rings were replaced by *trans*-1,4-substituted cyclohexane rings. Also, Hoffmann-La Roche (1973-74) developed[19] pyrimidine analogues and the group at Halle (1980)[20] dioxanyl analogues. As well as providing new and very useful materials to the electronics industry, these materials widened perception of structure/property relations. For example, it was now obvious that cyclohexane rings can very usefully give higher T_{N-I} values, lower viscosities and lower birefringence values than aromatic systems. Bicyclo(2.2.2.)octane analogues therefore became logical systems to study. Also, the stability of directly ring-ring linked systems opened minds to linking moieties which would not destabilise the materials, e.g., by the use of an inter-ring $-CH_2-CH_2-$ linkage, producing[21] the PECH and BIECH materials (1983).

This simple concept continues to be of value. For example, quite recently in an Alvey Research Programme (1985-88) on novel materials for ferroelectric LC displays, terphenyls such as

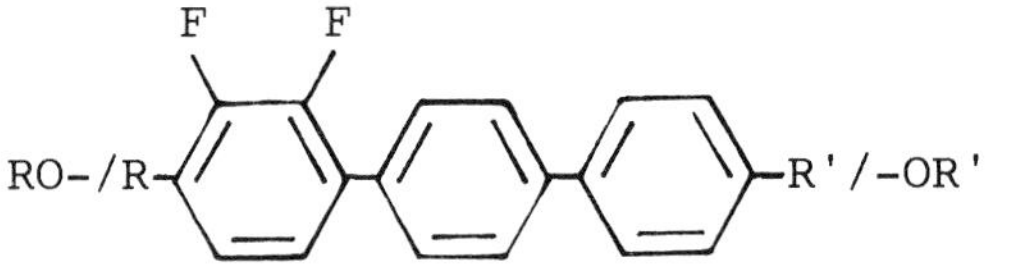

F F

RO-/R- -R'/-OR' and

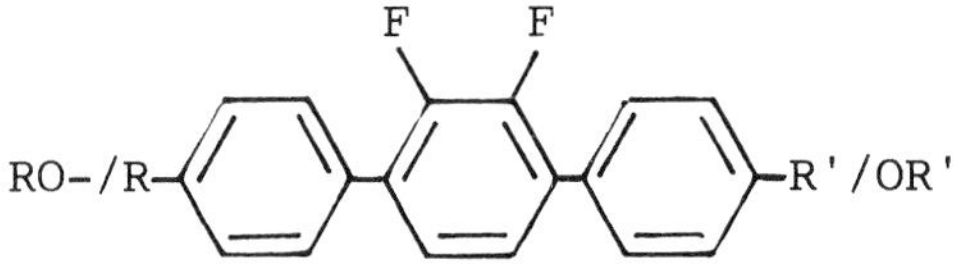

F F

RO-/R- -R'/OR'

were found to be very successful in providing host materials that can be doped with chiral cyanohydrin esters to give chiral smectic C materials capable of switching speeds as low as 7 μs. Such directly ring-ring linked materials cannot readily be accessed of course from the

HISTORICAL SEQUENCE OF NEMATOGEN DEVELOPMENT

date	nematogen	type	source
1972/73	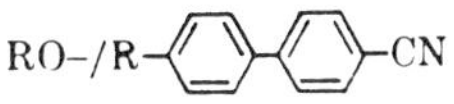 RO–/R– –CN	M/K	Hull; B.D.H.
	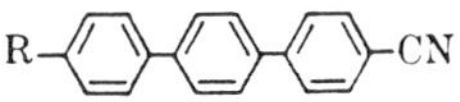 R– –CN	T	
1973	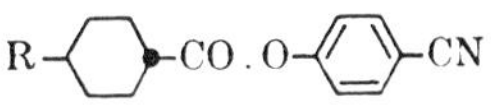 R– –CO.O– –CN	CH esters	Halle, D.D.R.
1973/74	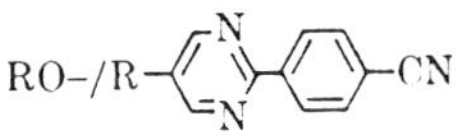 RO–/R– N N –CN	pyrimidines	Hoffmann-La Roche
1976/77	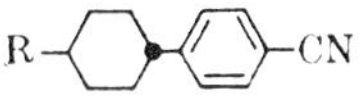 R– –CN	PCH	Merck, F.R.G.
	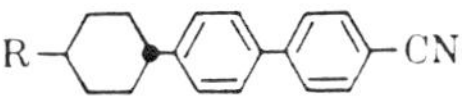 R– –CN	BICH	
	 R– –CN	CCH	
1980	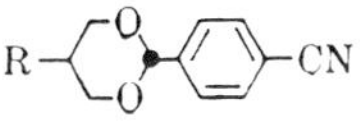 R– O O –CN	PDX	Halle, D.D.R.
1983	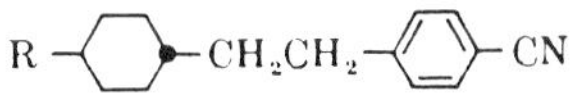 R – –CH_2CH_2– –CN	PECH	Hull; B.D.H.
	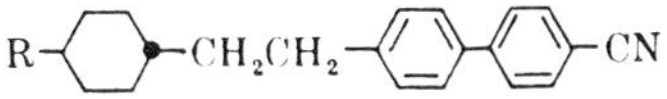 R– –CH_2CH_2– –CN	BIECH	

<u>ALSO</u>: BCO Materials (Hull/BDH)

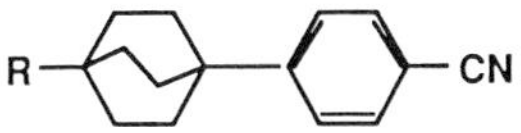

R– –CN

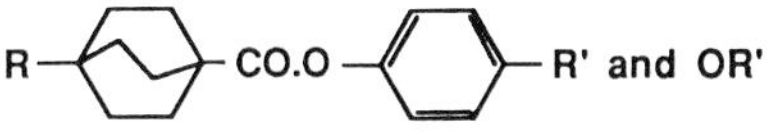

R– –CO.O– –R' and OR'

parent hydrocarbon *p*-terphenyl, and have to be built up from benzenoid and biphenyl precursors by carbon-carbon bond forming processes.

This can now be achieved very successfully by palladium catalysed boronic acid coupling processes such as:

(i) nBuLi, THF
(ii) $(^{i}PrO)_3B$, THF
(iii) 10% HCl
(i) and (ii) at -78 °C

$B(OH)_2$

Benzene, 2 M Na_2CO_3, $Pd(PPh_3)_4$

R—Br

R(O)—Br

(O)R

The scope and utility of such C-C coupling processes has therefore become much more widely recognised, and it has been shown that these bench synthetic procedures can very successfully be carried out on the large scale required by industry.

5 CONCLUDING COMMENTS

Looking at the guidelines for speakers, it is indicated that some attention should be paid to the human, social, economic and environmental aspects of the associated branch of the chemical industry. In this context, the following points may be made.

1. When commercialisation of the biphenyls/terphenyls was contemplated in the early 1970's, appropriate effort and funding were given to proving that the materials were non-toxic and non-carcinogenic. This was important, not only from the point of view of those manufacturing the materials, but also of those operating production lines for displays and of the community at large in the event of device breakage and the ultimate disposal of disused devices.

2. Rendering harmless cyano-containing waste products from the manufacturing process is dealt with efficiently.

3. The clear economic success of the work is obviously an all round beneficial aspect.

4. The devices made from the materials are energy saving - battery operated, at low voltages, with minimal power consumption.

5. The large area, flat panel displays present their data to the user in a kind-to-the-eye, flicker free format that is operator friendly.

Finally, it has been a privilege for me, for many years an academic, and now working in industry, to have played a role in this area that has been accorded milestone status by the RSC. The overall success of the work has been contributed to by timeliness, but has owed most to close collaboration of university chemists, industrial chemists, physicists, and electronic engineers, and for many years the positive outcomes of the endeavour have been hailed as a shining example of what can be achieved by University/Industry collaboration and the will to overcome the divisions between scientific disciplines as disparate as organic chemistry and electronic engineering. Putting different research groups into an interactive situation is not however an automatic guarantee of success, unless the right personalities and leadership are there. In this context, the project under review here was particularly well served and especial mention should be made of Professor C Hilsum,CBE,FRS (now at GEC Research), Dr J Kirton, and Professor E P Raynes,FRS all at RSRE, Malvern and of Dr B Sturgeon and Dr M G Pellatt at BDH Ltd (now Merck Ltd).

It is not the first time that this story has been told, and those interested to know more detail about it and the personalities involved are strongly recommended to see ref 3. It is encouraging to note that today, in 1991, the research groups and many of the personalities involved continue to work together on advanced aspects of the technological applications of liquid crystals such as electroclinic effects, liquid crystal polymers and polymer dispersed liquid crystal displays.

REFERENCES

1. G.W. Gray, K.J. Harrison and J.A. Nash, Electron. Lett., 1973, 9, 130.
2. F. Funada,'Fine Chemicals for the Electronics Industry' (ed. by D.J. Ando and M.G. Pellatt), Royal Society of Chemistry, Cambridge, 1991, p. 97.
3. C. Hilsum,'Technology of Chemicals and Materials for Electronics' (ed by E.R. Howells), Ellis Horwood, Chichester, 1984, Chap. 3.
4. G.W. Gray, Phil. Trans. R. Soc.,Lond.,1990, A330, 73.
5. H. Kelker and B. Scheurle, J. Phys., Paris, 1969, 30, 104; Angew. Chem. Int. Ed. Engl., 1969, 8, 884.
6. G.W. Gray, Phil. Trans. R. Soc., Lond.,1983, A309,77.
7. J. L. Fergason, U.S. Patent 3,731,986, 1973.
8. M. Schadt and W. Helfrich, Appl. Phys. Lett.,1971, 18, 127.
9. C.M. Waters, V. Brimmell and E.P. Raynes, Proc. 3rd Int. Display Res. Conf, Kobe, Japan, 1983, 396.
10. P. Bos and K. Koehler-Beran, Mol. Cryst. Liq. Cryst., 1984, 133, 329.
11. G.W. Gray, 'Advances in Liquid Crystal Materials for Applications', BDH Special Publication, 1978.
12. D. Coates, W.A. Crossland, J.H. Morrissy and B. Needham, J. Phys. D, Appl. Phys., 1978, 11, 2025.
13. G.W. Gray, J. Phys., Paris, 1975, 36, 365.
14. E.P. Raynes, 'Fine Chemicals for the Electronics Industry II' (ed. by D.J. Ando and M.G. Pellatt), Royal Society of Chemistry, Cambridge, 1991, p. 130
15. G.W. Gray and D.G. McDonnell, Electron. Lett., 1975, 11, 556.
16. I.C. Sage, 'Thermotropic Liquid Crystals' (ed. by G.W. Gray), CRAC Vol. 22, Wiley & Sons, Chichester, 1987, Chap.3.
17. R. Eidenschink, D. Erdmann, J. Krause and L. Pohl, Angew. Chem. Int. Ed. Engl., 1977, 16, 100.
18. R. Eidenschink, D. Erdmann, J. Krause and L. Pohl,

Angew. Chem. Int. Ed. Engl., 1978, 17, 133.
19. A. Boller, M. Cereghetti, M. Schadt and M. Scherrer, Mol. Cryst. Liq. Cryst., 1977, 42, 215.
20. D. Demus and H. Zaschke, Mol. Cryst. Liq. Cryst., 1981, 63, 129.
21. N. Carr, G.W. Gray and D.G. McDonnell, Mol. Cryst. Liq. Cryst., 1983, 97, 13.
22. G.W. Gray, M. Hird and K.J. Toyne, Mol. Cryst. Liq. Cryst., 1990, 191, 1.
23. G.W. Gray, M. Hird and K.J. Toyne, J. Chem. Soc. , Perkin Trans., 1989, 2041.

Subject Index